AIR POLLUTION
TECHNOLOGY

AIR POLLUTION TECHNOLOGY

DEAN E. PAINTER

Pitt Technical Institute
Greenville, North Carolina

Reston Publishing Company, Inc., Reston, Virginia 22090
A Prentice-Hall Company

Library of Congress Cataloging in Publication Data

Painter, Dean E.   1916–
Air pollution technology.
Includes bibliographies.
1. Air—Pollution. 2. Pollution control
equipment. I. Title. [DNLM: 1. Air pollution—
Prevention and control. WA754 P148a 1974]
TD883.P34   628.5'3   73–13949
ISBN 0–87909–009–X

© 1974 by

RESTON PUBLISHING COMPANY, INC.
A Prentice-Hall Company
Box 547
Reston, Virginia 22090

10 9 8 7 6 5 4 3 2 1

Printed in the United States of America.

ACKNOWLEDGEMENTS

I wish to thank the many individuals and organizations that have provided valuable assistance to me in preparing this textbook. Organizations that have provided photographs have been given credit lines for their kind assistance.

I wish to give special credit to the Environmental Protection Agency's Institute for Air Pollution Training, Office of Manpower Development, for providing me with the opportunity to attend several of their courses, allowing me permission to extract materials from their training manuals, and providing me with other assistance and best wishes from their highly qualified and dedicated staff.

I also wish to express my deep appreciation to my daughter, Mollie Painter, for editing and typing my early manuscripts; to my wife, Julia Painter, for her patience and encouragement through the many hours of research and writing; and to Lorraine Murphy for the typing of the early drafts of this book.

Finally, I wish to express my appreciation to Dr. Amos O. Clark for proofreading and to Jane Pridgeon, Jane Radford, and Katie Bryant for typing final drafts of the manuscript.

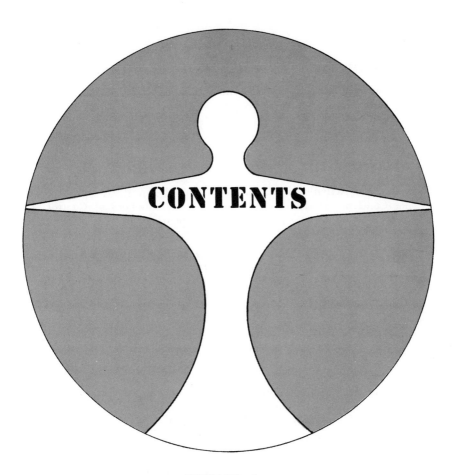

CONTENTS

PREFACE

In President Nixon's State of the Union message delivered on January 22, 1970 he posed a very real problem: "The great question of the '70s is—Shall we surrender to our surroundings, or shall we make our peace with nature and begin to make reparations for the damage we have done to our air, to our land and to our water?"

In our attempt to gain a higher standard of living we are constantly adding new and more complex dimensions to the air pollution problem. This does not mean that progress must stop in order to have a clean environment. Numerous air pollution disasters that occurred in the past might have been prevented had the public been more aware of and concerned with air pollution and with the means to abate air pollution. As the President of the United States has said— the foundation on which environmental progress rests in our society is a responsible and informed citizenry.

Beyond this general awareness and concern of all of our citizenry is the grave need for trained air pollution technicians to man the air pollution control agencies and to assist industry in establishing air pollution control measures that include sampling for pollutants, as well as testing, operation, and maintenance of air pollution control equipment.

A wide variety of pamphlets have been written concerning the problem of air pollution and the technology related to it, but, to date, no suitable textbook is available that presents this information in a concise and usable form. This text was written to fill that need.

Air Pollution Technology was written as a college level text. It is intended to support an introductory course for persons who intend to become involved in air pollution control activities or to support a university-level interdisciplinary program designed to make all students aware of their responsibilities.

For the interdisciplinary program some of the terminology used is rather technical in nature; however, technical terms have been explained in laymen's terms to reduce possible confusion. A glossary of terms is included in the appendix.

This text provides an introduction to the very technical science of air pollution technology. Other more detailed books are required to fully cover specific areas such as atmospheric and source sampling and analysis.

The format of the text introduces the reader to the air pollution problem by defining air pollution, by discussing the factors that contribute to air pollution, by listing types of pollutants and the principal sources of air pollution. The effects of air pollutants on man, animal, vegetation, and non-living materials are also covered.

The discussion of atmospheric sampling and analysis is followed by chapters on source sampling and analysis and source-emission inventories—these include descriptions of air quality criteria, air quality standards, and air pollution regulations. Chapters on control techniques for fixed and mobile sources are followed by a chapter on meteorological effects related to air pollution. The final chapter outlines an air resources management program based on the data presented in the preceding chapters.

References are listed at the end of each chapter to support the text material. Appropriate supporting films are listed. Questions and problems are presented following the chapters to emphasize the important issues and to aid the student in summarizing the material.

This text should be supported with laboratory field trips and the showing of training films and film strips for a better understanding of equipment shown in the text figures.

AIR POLLUTION
TECHNOLOGY

1

INTRODUCTION
TO
AIR POLLUTION

Student Objectives

—*To become aware of environmental problems.*
—*To learn what federal agency is responsible for organizing the fight against environmental problems.*
—*To learn the definition of air contaminants and air pollution and the factors that contribute to air pollution.*
—*To learn the difference between industrial air pollution and photochemical air pollution.*

Morris Neiburger, a noted meteorologist at the University of California, was quoted in *Today's Health*, an American Medical Association publication, as saying: "All civilization will pass away, not from a sudden cataclysm like a nuclear war, but from gradual suffocation in its own wastes." Some historians have ventured the supposition that some of the past great civilizations have probably done just exactly that.

1

The population explosion has led to a greater demand for more food, more water, more shelter, more transportation, and more manufactured goods. Satisfaction of these demands has led to a never-ending pollution of the air and sea around us and the earth beneath us.

The clear, sparkling rivers and lakes of our ancestors' time are fast becoming malodorous streams and putrid lakes devoid of life. The natural purification of these waterways, which in the past made it possible to reuse their water, is almost nonexistent today because of the extensive contamination by sewage, detergents, pesticides, and industrial wastes.

The airplane, diesel train, gasoline automobile, and the burning of fuel and refuse have all combined to make open sewers of our skies. To quote an article, "Man—An Endangered Species," from the *1968 Department of the Interior Yearbook*: "We have enhanced the future of everything except the overall future of the human race." Babies and bulldozers have been the sweet music of the progress of the past; possibly it is time to establish a new index for the progress of the future if man is to survive.

It is often stated that man-made pollution has three characteristics:

Pollution respects no political boundaries.

Pollution often arises from good intentions.

Pollution problems are created by society's misuse of technology.

To emphasize the solemnity of the pollution problem, another characteristic needs to be added:

It is difficult to convince the public that prevention is cheaper than correction!

Irving S. Bengelsdorf of the *Los Angeles Times* stated that since scientists and engineers are the most aware of our pollution problems, it is their responsibility to clarify public values, define policy options, and assist political leaders in the guidance and control of the powerful forces troubling our planet.

However, pollution problems cannot be solved by technology alone, since social and economic aspects must also be considered. The basic causes of our environmental troubles are complex and deeply embedded. We must substitute qualitative growth for quantitative growth, provide full accounting for the social costs of pollution, consider environmental factors in planning and decision making, and perceive the environment as a totality. We must understand and recognize the fundamental interdependence of all facets of the environment, including man himself.

On the national level on January 1, 1970, the National Environmental Policy Act of 1969 was enacted into law as Public Law No. 91–190. This Act established the Council on Environmental Quality and charged it with coordinating all environmental quality programs and with making a thorough review of all other federal programs that affect the environment.

On December 2, 1970, the Environmental Protection Agency (EPA) was established to unify the efforts of numerous agencies previously scattered throughout many federal departments and bureaus dealing with environmental problems. The mission of the EPA is to organize the fight against environmental problems on an integrated basis that acknowledges the critical relationships between pollutants, forms of pollution, and control techniques.

The Council focuses on what our broad policies in the environmental field should be, while the EPA focuses on setting and enforcing pollution control.

In the consolidation of environmental programs, state agencies continue to play a major role with the aid, guidance, and encouragement of the EPA.

Pollution of the air and sea is also an international problem, requiring a worldwide effort to provide solutions. The United Nations Educational, Scientific, and Cultural Organization took the first step toward solving the problem, when a conference covering international pollution issues was held in Stockholm, Sweden, in June 1972.

The fight against pollution does not belong to the scientists and engineers, the national government, and the international organizations alone; it must be recognized that each individual must also become personally involved if the fight is to be won.

President Nixon was quoted in the EPA publication *Toward a New Environmental Era* as saying: "The nineteen-seventies absolutely must be the years when America pays its debt to the past by reclaiming the purity of its air, its waters and our living environment. It is literally now or never."

DEFINITION OF AIR POLLUTION

air pollution

Air pollution is the presence in the outdoor atmosphere of one or more contaminants in a sufficient quantity and duration to cause them to be injurious to human health and welfare and animal and plant life and to interfere with the enjoyment of life and property.

air contaminants

Air contaminants include dust, fumes, mist, smoke, vapor, gas, or any combination of these. Air pollution consists principally of certain material by-products accompanying an energy-conversion process. These by-products can be referred to as *chemical* by their composition, *biological* by their interaction on living systems, and *economical* by their social characteristics.

Air pollution is a mixture of solids, liquids, and gases, which are dispersed rapidly by meteorological and topographic conditions. When pollutants become concentrated because of a lack of movement of air masses, air pollution disasters may occur. Also, otherwise harmless emissions into the atmosphere may become serious air pollutants when they combine with other materials present in the air.

FACTORS THAT CONTRIBUTE TO AIR POLLUTION

The factors that contribute to the creation of air pollution are both *natural* and *man–made*. The natural factors include meteorological and sometimes geographic conditions that restrict the normal dilution of contaminant emissions. Also, there are many naturally occurring airborne materials, such as pollen from flowers, gas from decaying matter, microorganisms (bacteria, virus, spores, molds), and particulates from nature's forest fires, dust storms, and volcanic eruptions.

Whereas the *natural factors* are usually beyond man's sphere of control, the *man-made factors* are more susceptible to control. Man-made pollutants resulting from heat and power generation (including nuclear power plants), solid-waste disposal, industrial processing, various types of transportation usage, nuclear warhead explosions, and other man-manipulated uses of radioactive isotopes are examples of these factors.

Man-made air pollutants are sometimes categorized generally as *industrial air pollution*, typical of the big cities where a great deal of coal and other fossil fuels are burned, or as *photochemical air pollution*, which results when the sun's rays combine with gaseous emissions to produce secondary air pollutants often related to weather inversions and high automobile density.

Radiation from exposure to cosmic rays as well as from radioactive gases released from soils and rocks has been with man from the beginning. However, in the past 70 to 80 years, man has added man-made radioactive pollutants to a degree that has approximately doubled the amount of radiation pollution in the atmosphere. As more use is made of nuclear energy for power plants and propulsion and of radioisotopes in medicine and scientific research, the radiation pollution problem will increase.

This text will concentrate mainly on the man-made air pollution that is not radioactive in nature, leaving the more complex field of radioactive pollution to advanced study.

REFERENCES

Council on Environmental Quality, *Environmental Quality—The First Annual Report.* Washington, D.C.: U.S. Government Printing Office, 1970.

U.S. Department of Health, Education, and Welfare, *Air Pollution Control Field Operations Manual*, PHS, Pub. No. 937. Washington, D.C.: U.S. Government Printing Office, 1962.

U.S. Environmental Protection Agency, *Federal Register*, National Ambient A Q Standards, Vol. 36, No. 67, pp. 6680–6701. Washington, D.C.: U.S. Government Printing Office, Apr. 7, 1971.

RECOMMENDED FILMS

M-1419X   Air of Disaster (50 min)
M-1418X   The Poisoned Air (50 min)
      Available: Distribution Branch
                National Audio-Visual Center (GSA)
                Washington, D.C.
                20409

QUESTIONS

1/ What are some of the characteristics of man-made pollution?

2/ What two professions are most aware of the air pollution problems?

3/ What federal agency is most responsible for dealing with environmental problems?

4/ What is air pollution?

5/ What are some of the factors that contribute to air pollution?

6/ What are the two major categories of man-made pollutants?

THE AIR POLLUTION PROBLEM

2

Student Objectives

—*To learn the meaning of air pollution abatement.*
—*To understand air contaminant classification by origin
and by physical state.*
—*To be able to recall two basic types of particulate pollutants, six inorganic compounds of most concern and two
organic compounds, and to give an example of each.*
—*To learn the four principal sources of air pollution and
two types of source classification used in source inventories.*

Reduction of existing pollution and prevention of future pollution is, with some exceptions, not generally a very difficult technological problem. However, pollution abatement is sometimes unsavory
from an economic point of view. It is important to realize that the
abundant life with continued economic growth has brought many

7

pollution problems that cannot be dealt with by passively hoping that technological knowledge alone will solve them. Society must be willing to contribute in precluding contamination of our earth.

To solve the air pollution problem, we must determine the type or quality of air that is most desirable. The types of air contaminants, their source of emission, and their ill effects need to be identified first. Then the control techniques for abatement and prevention of air pollutants need to be introduced.

TYPES OF AIR CONTAMINANTS

Under normal air conditions the atmosphere has been found to contain the following percentages of gases:

Oxygen (O_2)	20.94%
Nitrogen (N_2)	78.09%
Argon (Ar)	0.93%
Carbon dioxide (CO_2)	0.03%
Other	0.01%
	100.00%

These are strictly hypothetical percentages, since man and nature have injected many contaminants into the atmosphere to add to or displace the normal atmospheric gases. Several of these injected contaminants cause air pollution. The extent of their effect is determined by the amount of each compound released into the atmosphere and sometimes by a combination of a given compound with another compound.

Air contaminants may be classified on the basis of their *origin* or on their *physical state*.

origin

If pollutants are *emitted* directly into the atmosphere, such as sulfur oxides released by burning fossil fuels (i.e., coal and oil), they are classified as *primary* pollutants.

If pollutants are *formed* after emission into the atmosphere as a result of some reaction with matter already in the atmosphere, the resultant pollutant is classified as a *secondary* pollutant.

Most secondary pollutants result from *photochemical reactions* that utilize energy from the sun's rays. As an example, oxygen atoms may be split off from nitrogen oxides to combine with oxygen in the atmosphere to form ozone (O_3), a secondary pollutant. However, other *nonphoto-chemical reactions* may also produce secondary pollutants. For instance, *hydrolysis,* a chemical process of decomposition involving the splitting of a chemical bond and addition of water, can create secondary pollutants. As an example, sulfur oxides may combine with water in the atmosphere to form sulfuric acid (H_2SO_4), a very corrosive secondary pollutant. *Catalytic oxidation,* a chemical process wherein electrons are removed from a molecule, aided by a substance that itself is not altered in the reaction, may also produce secondary pollutants. An example of this is the catalytic oxidation of sulfur oxides after being adsorbed on the surface of suspended solid particles. These particulate sulfates are often responsible for reduced visibility. It is not possible to classify *every* pollutant discussed in the next section as to whether it is *always* a primary or secondary pollutant. Some matter emitted into the atmosphere may be a primary pollutant in its emitted form and may then undergo one of the reactions and become a secondary pollutant in another form.

physical state

Air contaminants are classified by their physical state into *particulate* pollutants (minute fragments of matter in solid or liquid form) or as *gaseous* pollutants. Gaseous pollutants may be further divided according to chemical composition as *inorganic* or *organic* gases.

Particulate pollutants have been described by the *Air Quality Criteria Pamphlet* (Particulates) as finely divided solid or liquid particles larger than a single small molecule (about 0.0002 micron in diameter) but smaller than about 500 microns. [One micron (μ) is 1/1,000 millimeter (mm) or 1/1,000,000 of a meter. A millimeter is approximately 0.04 in. One micron (μ) is the same as one micrometer (μm) and the terms may be used interchangeably. Particles in this size range have a lifetime in the suspended state varying from a few seconds to several months.

Fumes and smoke formed as combustion products and photochemical aerosols make up a large fraction of the particulates in the range from 0.1 to 1 μ diameter. Because of their small size, these particles tend to remain suspended in a free-floating state for long periods of time, and are broadly termed *suspended particulates*. Some examples are metallic fumes, and droplets of oil, tar, or acid.

Particles between 1 μ and 10 μ generally include soil, fine dusts, and soot emitted by industry and, at maritime locations, airborne sea salt. Industrial sources include municipal incinerators, cement plants, steel mills, sulfuric acid factories, and kraft pulp mills.

Particles larger than 10 μ frequently result from mechanical processes such as highway construction, wind erosion, grinding, spraying, and pulverizing of material by vehicles.

Particles larger than 1 μ tend to settle out of the atmosphere, due to gravitational pull because of their size, and are broadly referred to as *settleable particulates.*

Gaseous pollutants are pollutants present in the atmosphere in the physical state of a gas rather than as a solid or liquid. These gaseous pollutants emitted from various sources become scattered or mixed in with the atmospheric gases that are present under normal air conditions. They tend to remain suspended and do not settle out readily. The sources of the gaseous pollutants of greatest concern are discussed next.

Gaseous pollutants may also be classified according to chemical composition as *inorganic* or *organic.*

1/ *Inorganic gases* are composed of matter other than plant or animal in origin. Except for the very simple carbon compounds, carbon monoxide (CO) and carbon dioxide (CO_2), these inorganic gases are non-carbon-containing compounds. Those inorganic gases of most concern are

a/ *Sulfur compounds.* The sulfur oxides (SO_2, SO_3) and hydrogen sulfide (H_2S) are mostly primary in origin and result from the burning of coal, oil, and diesel fuels and the processing of petroleum, chemicals, metals, and minerals. Hydrogen sulfide is a major odor-producing pollutant. An example of the production of sulfur compounds as secondary pollutants was previously covered.

b/ *Nitrogen compounds.* The nitrogen oxides (NO, NO_2) are primary pollutants released by petroleum operations and by industrial and automobile combustion. Ammonia (NH_3) is produced by fuel combustion and chemical production as a primary pollutant. NO and NO_2 may combine photochemically with other gases under certain conditions to produce secondary pollutants. An example is the production of ozone (O_3) referred to previously.

c/ *Chloride compounds.* The most common primary pollutant forms are chlorine gas and hydrogen chloride gas produced from cotton- and flour-bleaching processes and petroleum refining. A secondary pollutant, hydrochloric acid, may be produced by hydrolysis of hydrogen chloride gas.

d/ *Fluoride compounds.* This active nonmetal produces primary pollutants of silicone tetrafluoride (SiF_4) and gaseous hydrogen fluoride (HF). These emissions are principally from petroleum refineries, fertilizer factories, aluminum producers, steel plants, pottery works, and brick plants. A secondary pollutant of fluoride mist may be formed by hydrolysis of the preceding gases.

e/ *Carbon compounds.* Carbon monoxide (CO) is a primary pollutant produced by incomplete combustion of gasoline and to a lesser extent by the fuming of metal oxides. Carbon dioxide (CO_2) is also produced from these sources, but is not usually considered a pollutant; however, CO and CO_2 play an important part in the production of secondary pollutants of carbon compounds.

f/ *Oxidants.* Ozone and nitrogen dioxide are the major inorganic oxidants in the atmosphere. Ozone is formed naturally in the atmosphere by electrical discharges, but usually does not reach pollutant proportions unless nitrogen dioxide coupled with the ultraviolet rays of the sun reacts with oxygen and other molecules, particularly organic and hydrocarbons to stabilize the ozone as a secondary pollutant. The nitrogen oxides were previously described as primary pollutants.

2/ *Organic gases* are composed of matter mostly derived from living organisms, compounds that contain carbon and hydrogen and may contain other elements. Organic compounds may be further defined as compounds with covalent bonding or sharing of electrons and compounds with carbon–carbon linkage or carbon–hydrogen linkage. Those organic gases of most concern are

a/ *Hydrocarbons* (HC) which are the organic gases containing only hydrogen and carbon. The principal source of these primary pollutants is petroleum production and the inefficient combustion of fuels by gasoline and diesel-powered vehicles, gas-turbined and jet aircraft, and solvent usage.

b/ *Derivatives of hydrocarbons* which may be released as primary pollutants or they may be secondary pollutants formed in the atmosphere as the result of certain photochemical reactions. They may be *oxygenated* hydrocarbons (HC derived from combination with oxygen) such as aldehydes and acrolein, which are by-products of fat decomposition and surface coating. They may be *halogenated* hydrocarbons (HC derived from combination with fluorine, chlorine, bromine, and iodine), such as carbon tetrachloride, that are by-products of degreasing, dry cleaning, and the use of solvents.

PRINCIPAL SOURCES OF AIR POLLUTION

On a tonnage basis, the EPA's data on national emissions of major air pollutants indicate that the four principal sources of air pollution are transportation, fuel combustion from stationary sources, indusrial processes, and solid-waste disposal.

Although transportation includes ships, planes, trains, and automobiles, the automobile is the worst contributor to air pollution. Efforts of the automotive industries to control these emissions are covered in Chapter 11.

Central power station boilers are the major contributor to the second source, fuel combustion from stationary sources. Our increase in population continues to generate greater demands for more power. Only through conversion to higher-grade fuels with less pollution potential can we reduce this problem.

Process industries are industries that process raw materials into useful products. Examples are smelters, steel mills, petroleum refineries, producers of rubber products, textiles, paper, and chemicals. These industries produce specific process pollutants in addition to the general pollutants produced by the combustion of fuel to produce energy for operation of the processing plant.

Solid-waste disposal produces air pollution from open burning dumps, incinerators, and improperly operated sanitary landfills. Air pollution laws against open burning dumps should greatly decrease this major source of air pollution in the future.

specific source

Industries are classified as *specific sources* due to the fact that each industry presents unique problems, depending upon what the manufacturing operation involves, for example, raw materials, fuels, process method, operating efficiency, and pollution control devices installed. Industries occupy a limited area relative to communities, thus permitting a source-by-source evaluation of their pollution potential. Table 2–1 lists the standard industrial categories, a brief description of their activities, and the types of air pollution problems involved in

TABLE 2-1 Standard Industrial Categories (Specific Source Classification)

Manufacturing Industry	Nature of Activity	Types of Air Pollution Problem
Primary metals (ferrous and nonferrous)	Primary smelting of ore to obtain metallic elements. Steel mills—manufacture of steel alloy products by removal of graphitic carbon from iron and addition of alloy elements. Ferrous and nonferrous foundries—cast products from sand or permanent molds. Secondary smelting—separates ingots of each element from scrap. Secondary ingot production—prepares alloyed ingots from scrap.	Primarily fuming of metallic oxides, and emission of CO, smoke, dust and ash from melting operation, depending on the volatility and impurities of the metals, scrap or ore concentration. Smelting is most notorious, emitting sulfur dioxide, lead and arsenical copper fume, depending on metal smelted.
Fabricated metal products	Manufacture of a large variety of products: Heating and plumbing equipment, tools and hardware, structural metal products, cutlery, metal stamping and coating, lighting fixtures, tin cans and others. Usually involves metal melting from ingot; machine shops, metal finishing and surface coating.	Metals melted are usually refined, and melting operations are easily controlled. Principal air contaminants are metallic fumes and dusts from foundries and solvent mists and vapors from application of protective coatings in finishing departments.
Machinery	Machining and finishing of component machinery parts and/or their assembly in the production of a wide variety of mechanical equipment (but not including electrical machinery). Farm implements, machine tools, printing, office and store equipment, oil field production and refinery equipment,	Primarily dusts and mists from finishing departments, some smoke and fumes from quenching in tempering and heat treating. Metal melting is not usually involved.

13

TABLE 2-1   (Cont'd)

Manufacturing Industry	Nature of Activity	Types of Air Pollution Problem
	textile, shoes and clothing equipment, construction equipment, household equipment, etc.	
Electrical machinery	Manufacturing and assembly of machinery; apparatus and supplies for the generation, storage, transmission, and utilization of electrical energy, principally electrical motors and generators.	Air contaminants similar to those described under machinery.
Mining	Quarrying and milling of solid products and minerals—coal, iron and metallic ore.	Waste explosive gases, CO, etc., dusts and fumes.
	Petroleum and petroleum refining. Drilling and extraction of crude petroleum from oil fields, recovery of oil from oil sands and oil shale, and production of natural gasoline and cycle condensate. Oil refining consists of a number of complex flow processes based on heat and pressure which crack, build up, alter or segregate hydrocarbons from crude oil in the production of a large variety of commercial products from high octane gasolines to heavy oils and greases. Natural gas originates from the oil fields in the southwest.	Due to the large number of production steps, all forms of air pollution arise from refineries. These include vapors from evaporation of petroleum products in handling and storage; sulfur dioxide and smoke plumes from scavenging and burning of refinery fuels in heating equipment; odors, mists and dusts from cracking operations.
Furniture, lumber and wood products	Logging and milling, including veneering, planing, and plywood manufacturing; boxing and container manufacturing; sawdust and other by-product manufacturing. Furniture mfg., household, office	Fines and dusts from milling operations. Paint and solvent emissions from surface coating. Smoke from burning waste lumber, mill ends, fines and sawdust.

Industry	Description	Air pollution effects
	and store fixtures. Involves production wood working (planing, milling, cutting, sanding, shaping, etc.), finishing (staining, priming, painting, etc.) and occasionally elimination of large volume production wastes by burning.	
Transportation equipment	Manufacture and/or assembly of component parts for ships, automobiles, rolling stock, aircraft and other transportation equipment involving fabrication of structural assemblies and components, and, in the case of ships and rolling stock, riveting, welding and sheet metal work. A high degree of specialization, especially in automobiles and aircraft, necessitates extensive subcontracting activities, or concentration of many captive industries into coordinated production systems.	Aside from assembly lines which are not in themselves significant sources of air pollution, captive subsidiary operations may involve foundries, heat treating, wood-working, plating, anodizing, chem-milling and surface coating operations which contribute all types of air contaminants including organic vapor emissions from the application, drying and baking of protective coatings.
Chemicals and allied products	Manufacture of almost an unlimited variety of products: petro-chemicals, heavy or industrial chemicals such as sulfuric acid, soda ash, caustic soda, chlorine and ammonia; pharmaceuticals, pesticides, products of nuclear fission, plastics, cosmetics, soaps, synthetic fibers, such as nylon, pigments, etc. Manufacturing techniques encompass virtually the entire chemical technology.	Chemical technology makes possible all forms of pollution, involving the emissions of the chemicals (both chemical and end-product) and the derivative or reaction products of the chemicals in process or in the atmosphere.
Minerals (stone, clay and glass products)	Manufacture from earth materials (stone, clay and sand), glass, cement, clay products, pottery, concrete and gypsum products, cut stone products, abrasive and asbestos products, roofing materials,	Dusts from mechanical processes, smoke and fumes from melting or kiln operations.

TABLE 2-1 (Cont'd)

Manufacturing Industry	Nature of Activity	Types of Air Pollution Problem
	bricks, etc., involving mechanical processes such as crushing, mixing, classifying and grading; batching, drying and baking in kilns to vitrify dishware, and melting and forming to produce glass products.	
Textile	Includes milling and manufacturing of yarns, threads, braids, twines, fabrics, rugs, apparel, lace, and a vast variety of products involving processes of spinning, spooling, winding, weaving, braiding, knitting, sewing, bleaching, dyeing, printing, impregnating, batting, padding, etc.	Lint and fines are emitted from production wastes; organic vapor emissions or other mists from dyeing, bleaching, impregnating, cleaning; smoke from combustion equipment required to power weaves, looms, and other processing and conveyor equipment.
Rubber products	Manufacture from natural, synthetic, or re-claimed rubber (gutta percha, balata, or gutta siak), rubber products such as tires, rubber footwear, mechanical rubber goods, heels and soles, flooring, and other rubber products. Processes involve mastication, mixing or blending of crude rubber, reclaim or chemical rubbers, calendaring, tubing, binding and cementing, curing, etc.	Local dusts and carbon black emissions from mixing and rolling operations, but usually under careful control. Organic vapor emissions from solvents used in bonding and cementing, coating and drying of products.
Paper and allied products	Manufacture of paper and paper products from wood pulp, cellulose fibers, and rags involving cutting, crushing, mixing, cooking, and paper mills.	Some possible sawdust emissions, but other-wise practically no emissions, except from combustion equipment to provide steam heat and power for mechanical equipment. Construction materials such as roofing paper

involve saturating paper with asphalt and impregnating with minerals, causing mist and dust problems.

Category	Description	Emissions
Printing and publishing	Printing and publishing by means of letterpress, lithography, gravure, or screen, bookbinding, typesetting, engraving, photo-engraving, and electrotyping. Involves lead melting pots for typesetting machines, and significant quantities of inks containing organic solvents.	Lead oxide emissions are possible from lead pots, but these are easily controlled. Organic solvent emissions arise from the large volume of inks, particularly in rotogravure processes.
Instruments	Manufacture and assembly of mechanical, electrical and chemical instruments for dental, laboratory, research and photographic uses, including watches and clocks. Involves casting and machining of a variety of hard metal alloys, including brass and steel; assembly, plating and finishing.	Emissions from these plants are usually controlled, but can involve smokes, dusts, and fumes similar to those of fabricating and machinery manufacturing industries. Hardchrome electrolytic plating is usually involved with high quality instrumentation, causing emission of acid mists.
Food and kindred products	Includes the slaughtering of animals and the curing and smoking of meat products as well as the preparation of all other foods such as dairy products, canning and preserving of fruits, vegetables and seafoods; grain and feed milling, baking, preparation of beverages, including coffee, beer and other alcohols; animal rendering, manufacture of fats, oil, grease, tallow, etc.	Most notably odors, particularly from rendering operations and from poor housekeeping where products are permitted to decompose. Odors may also occur from the handling of by-products, and from coffee roasting. Dust from grain and feed mill operations.
Other manufacturing industries	Tobacco, ordnance and armaments, leather and leather products, building construction, jewelry and silverware, etc.	All types of air pollution arising from basic processes described in the foregoing.

each category. This table was extracted from the *Standard Industrial Classification Manual* listed in the references.

multiple source

When a source-by-source evaluation of each pollutant is not possible, a multiple-source classification is used. Table 2–2 lists examples of multiple sources.

TABLE 2-2   Examples of Multiple Source Classification

Odor Producing Activities	*Combustion of Fuels*	*Evaporation of Petroleum Products*	*Incineration of Solid Wastes*
Sewage	Mobile Ships	Storage	Private
Animals	Aircraft Railways	Marketing	Municipal
Refineries	Automobiles	Solvent use	Industrial

Combustible Wastes
Home incinerations
City incinerations
Open dump fires
	Stationary
	Homes
	Power production
	Service industries,
	  schools, motels,
	  hospitals, etc.
			Commercial

Industrial Processes Smelting Fertilizing Dry-cleaning Asphalt paving			
Combustion Processes Coke ovens Engine exhausts Heating systems			
Processing Plants Meat, fish, poultry			

source inventory

One of the early steps in solving air pollution problems is to locate the source from which air contaminants are being emitted. The air pollution control agency locates, enumerates, tabulates, and classifies pollution sources according to the quantity and quality of the materials processed. The specific-source or multiple-source classifications listed in the Tables 2–1 and 2–2 are used in these source inventories. The steps in compiling source inventory and emission inventories are covered in Chapter 9.

REFERENCES

U.S. Bureau of the Budget, *Standard Industrial Classification Manual.* Washington, D.C.: U.S. Government Printing Office, 1957 and 1958.

U.S. Department of Health, Education, and Welfare, *Air Quality Criteria Pamphlet* (Particulates), PHS (NAPCA), Pub. No. AP-49. Washington, D.C.: U.S. Government Printing Office, 1969.

RECOMMENDED FILMS

MIS-679   Breathe at Your Own Risk (58 min)
MIS-677   Sources of Air Pollution (5 min)
Available:
   Distribution Branch
   National Audio-Visual Center
   (GSA)
   Washington, D.C.
   20409

QUESTIONS

1/ What is pollution abatement?
2/ What are the four principal sources of air pollution?
3/ What is the difference between a multiple-source and a specific-source classification? Why are they treated differently?
4/ List three examples of a multiple-source classification.

5/ Air contaminants are divided into two types. What are they?

6/ What is the difference between a primary and a secondary pollutant?

7/ What is a particulate air pollutant?

8/ What is the difference between suspended and settleable particulates?

9/ List five inorganic gaseous pollutants of most concern.

10/ What are the major organic gaseous pollutants?

3

EFFECTS OF
AIR POLLUTANTS ON
MAN

Student Objectives

—*To gain an understanding of the sources of information available upon which effects of air pollution on man are based.*
—*To understand the basic functions of the respiratory system and how it is affected by air pollution.*
—*To become aware of the various types of pollutants and their sources that are of concern in effects on man.*
—*To relate the effects of these pollutants to specific diseases of man.*

BASIS OF FACTS

In 1962 at the first national conference on air pollution, the U.S. Surgeon General stated that our knowledge of the health effects of air

21

pollution has been amplified through three types of investigations: *statistical studies* cross sampling past illnesses and deaths, *epidemiological studies* reporting death and respiratory diseases, and *laboratory studies* listing animal and human responses associated with air pollution.

statistical studies of air pollution episodes

Certain disasters show the dramatic nature of air pollution. *In all these disaster cases a temperature inversion sealed in the air space of a community and caused serious illness, discomfort, and excess death (more than would have occurred under normal conditions).* In December 1930, a thick, stagnant fog enveloped a heavily industrialized area of the Meuse Valley, Belgium. Sixty-three persons died and 6,000 persons became ill with throat irritations, hoarseness, coughing, and breathlessness. In October 1948, a similar fog blanketed Donora, Pennsylvania, where, over a period of 4 days, 20 died and 6,000 became ill with coughs, sore throats, headaches, burning of the eyes, nausea, and nasal discharges. In Poza Rica, Mexico, in 1950, toxic hydrogen sulfide was released inadvertently from a plant recovering sulfur from natural gas. Three hundred were hospitalized and 22 died from respiratory and central nervous system disorders. An inversion existed at the time. In December 1952, London experienced a pea-soup fog for 4 days; 1,600 more deaths were reported than would have normally occurred. Cardiorespiratory illnesses increased sharply over a 7-day period that began with the first day of the fog. In 1953, 1962, and again in 1966, New York City went on a 24-hour disaster alert because of dangerous pollution situations.

Air pollution related to health is basically an *ecological problem.* It is a problem not only for humans but for other forms of animal life as well as plant life. Most air pollutants, like other forms of matter, in time appear in water, or in soil, and eventually in human food.

Little is known about the long-range effects of air pollution on health; therefore, caution must be used in stating that air pollution *causes* certain diseases. However, according to studies by the U.S. Department of Health, Education, and Welfare (HEW), there is strong evidence that air pollution accompanied by prolonged stagnation aggravates and contributes to respiratory ailments such as chronic bronchitis, pulmonary emphysema, bronchial asthma, lung cancer, and pulmonary edema.

Specifically, some results of studies by HEW show that death rates from cardiorespiratory diseases correlate with air pollution levels (e.g.,

Meuse, Donora, London). Asthmatic attacks among susceptible patients correlated with variations in sulfate air pollution, and these attacks led to employee absenteeism during episodes that occurred in Nashville and New Orleans. Asthma attacks also correlate with burning of refuse. Air pollution episodes have dramatically pointed up that air pollution by particulates combined with SO_2, as well as sulfate, leads to high mortality and morbidity rates due to cardiorespiratory problems. Another well-known respiratory disease, lung cancer, develops when laboratory animals injected with flu virus are later subjected to ozonized gas vapors. The effects of ordinary concentrations of air pollutants are subtle, although health damage is real. A death certificate never shows air pollution as a cause of death because its effects are difficult to pinpoint. Statistics do indicate that urban areas with the highest air pollution content have more cases of asthma, bronchitis, and emphysema. There is no longer any doubt that air pollution is a hazard to health; evidence very much supports this supposition.

Population surveys have been designed to measure the prevalence of chronic nonspecific respiratory diseases and to relate these measures to the degree of air pollution in specific areas. Questionnaires aimed at disclosing the frequency and duration of coughs, production of phlegm (viscid mucous secreted in abnormal quantities in respiratory passages) and sputum (saliva discharged from respiratory passages), shortness of breath, and occurrence of deep chest illness have been used to define chronic respiratory diseases.

epidemiological studies

Epidemiological studies (the systematic study of naturally occurring associations) usually involve *an extensive study of a densely populated area* such as Los Angeles and the related effects of smog on the population; *an occupational approach related to a specific industry,* such as brown lung in the cotton-mill industry; *personal pollution* related to cigarette smoking; or *natural pollution* related to pollen and its effect on asthma or hay-fever patients. Epidemiological studies show a close correlation between sulfur dioxide and particulate pollution and severe health effects during air pollution episodes.

laboratory studies

Laboratory studies (clinical studies) can be used to sort out specific pollutants and to determine the effects of specific dosages. However, the conditions that actually exist in an uncontrolled atmosphere are virtually impossible to duplicate in an artificial environment. Results of laboratory experiments, although helpful in establishing certain parameters of the air pollution problem, can be dangerous if accepted at face value without taking into account their simulated setting.

Public Health Service Publication No. AP-49, *Air Quality Criteria Pamphlet* (Particulates), states that laboratory studies do not provide sufficient data upon which to base air pollution criteria for particulates. Laboratory studies provide too much control by using young healthy animals, short-term exposures, constancy of population, constancy of exposure, and constancy of temperature and humidity.

TARGET ORGAN SYSTEMS

Man's health may be affected by air pollution indirectly by ingestion of foods contaminated by air pollution or directly by inspiration of gases and particulates through the human respiratory system. However, scientists engaged in studies designed to pinpoint relationships between air pollution and disease have concentrated on two human organ systems—the eye and the respiratory system, since these two systems are most affected by air pollution.

the eye

Although no lasting damage to the eye has been attributed to air pollution, studies show a reduction in sharpness of vision from excess carbon monoxide. Indirectly, smog, in its reduction of visibility, can be considered an eye irritant. Eye irritation may result when gaseous or particulate materials contact the external coat of the eye and the internal mucous lining of the eyelid. This contact may cause physical damage resulting from excessive rubbing of the eyes to relieve the irritation.

Atmospheric substances causing eye irritation have not been completely defined. However, some particulate materials combined with ozone, oxides of nitrogen, aromatic hydrocarbons, and synthetic pollutants do cause eye irritation.

Oxidant concentrations in photochemical smog exceeding 200 $\mu g/m^3$ will cause eye irritation and 250 $\mu g/m^3$ will aggravate respiratory diseases.* Ozone by itself, at expected atmospheric levels, is not an eye irritant, but the mixtures containing photochemical oxidants do correlate with eye irritation. The precursors (substances from which other substances are formed) of the eye irritants are organic compounds in combination with oxides of nitrogen, the most potent being aromatic hydrocarbons. The effective irritants in synthetic systems are formaldehyde, peroxybenzoyl nitrate (PB$_2$N), peroxyacetyl nitrate (PAN), and acrolein.

the respiratory system

The basic functions of the respiratory system (Fig. 3–1) are to inhale air into the lungs, filter impurities from the inspired air, supply oxygen contained in this inspired air to the circulatory system, and exhale carbon dioxide removed from the circulatory system.

The upper respiratory tract, consisting of the nasal cavity, nasal pharynx, larynx, and trachea, removes most particulates larger than 10 μ (microns) in diameter by inhaling and then immediately exhaling the air. Gaseous pollutants and particulates not exhaled are removed when they contact the mucous lining. This mucous blanket, aided by the projecting hair-like cilia, moves the particulates toward the pharynx, where they are swallowed, expectorated, or expelled by the cough reflex located in the larynx.

The lower respiratory tract consists of the bronchi, bronchioles, alveolar ducts, alveolar sacs, and the alveoli of the lungs. Air is first drawn through the large bronchi and then through a system of branching ducts, decreasing in size from the conducting bronchioles to the terminal bronchioles to the respiratory bronchioles. Further branching continues into the alveolar ducts, the alveolar sacs, and finally the

*In the past most air pollutants have been measured and reported as parts per million (ppm) or, for those found in very minute amounts, in parts per billion (ppb). More recently, pollutants have been reported in micrograms per cubic meter (μg/m³) or, in the case of carbon monoxide, in milligrams per cubic meter (mg/m³). (Carbon monoxide is usually found in the atmosphere in higher concentration than other gaseous pollutants.) In this textbook both means of expression are used. Since each type of pollutant varies in composition, a table for conversion of the most common pollutants from parts per million to micrograms per cubic meter is included in Chapter 7.

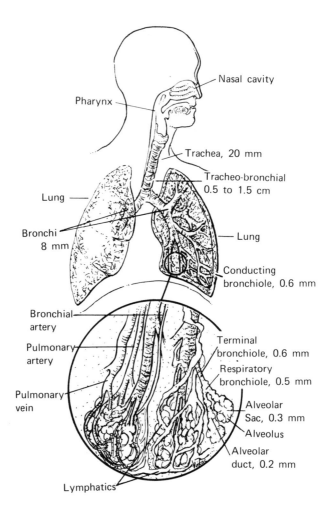

Nasal cavity

Pharynx

Trachea, 20 mm

Tracheo-bronchial
0.5 to 1.5 cm

Lung

Bronchi
8 mm

Lung

Conducting
bronchiole, 0.6 mm

Bronchial
artery

Pulmonary
artery

Terminal
bronchiole, 0.6 mm

Respiratory
bronchiole, 0.5 mm

Pulmonary
vein

Alveolar
Sac, 0.3 mm

Alveolus

Alveolar
duct, 0.2 mm

Lymphatics

Fig. 3–1 The human respiratory system.

alveoli. The alveoli are very minute structures surrounded by tiny capillary blood vessels.

Human respiration may be divided into *ventilation* and *respiration*. Ventilation is the mechanical process of inhaling ambient air into the alveoli of the lungs and exhaling waste air out of the lungs. Respiration is the exchange of gases between the alveoli and the blood. Respiration is further divided into *external respiration* and *internal respiration*.

External respiration is the gas exchange across the respiratory membrane of the lungs located in the alveolar ducts, alveoli, and alveolar sacs, and into the blood in the pulmonary capillaries (minute vessels of the circulatory system). This exchange of gases in the lungs is accelerated by the ventilation movement of the lungs, which speeds up diffusion of the gases by regulation of pressure. Gases diffuse rapidly from a point of higher pressure to points of lower pressure. The carbon dioxide diffuses from the site of high pressure in the lung capillaries to the point of low pressure in the alveoli, while oxygen diffuses from the alveoli into the capillaries.

Internal respiration is the diffusion of gases between the blood of the systemic capillaries and the cells of the body. After oxygen has diffused into the capillaries, it combines with the hemoglobin of the blood and is transported by the circulatory system in the solid form of iron oxide. This iron oxide reverts back into gaseous oxygen when it diffuses from the capillaries through the cell-wall membrane. At the cell wall an exchange is made, with carbon dioxide diffusing into the capillaries and oxygen diffusing from the capillaries and entering the cell. When the carbon dioxide from the cell diffuses through the cell membrane, it is converted into a solid form similar to baking soda. This solid material is transported by the circulatory system to the lungs, and when it reaches the lungs it is converted back into its original carbon dioxide gaseous state. The carbon dioxide gas is then expelled from the body on exhalation. (Figure 3–2 depicts transport of O_2-CO_2.)

Particulates less than 10 μ in diameter may enter the deepest areas of the lungs; however, those less than 1 μ in diameter are the

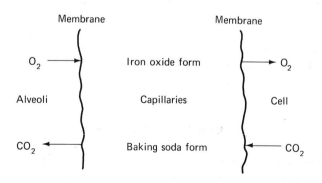

Fig. 3–2 Transport of O_2 to CO_2.

most likely to be retained. There are various ways in which these particles may be removed: by the mucous blanket beginning at the terminal bronchioles and extending upward into the upper respiratory tract; by body fluids that carry the particles to the lymph nodes where they are filtered out; by entering the bloodstream and being excreted through the kidneys; or by reacting with the lung tissues. The mucous membrane may become irritated by particulates, leading to respiratory difficulty.

When the particles react with lung tissues, three major types of lung damage may ensue: bronchitis—a congestion of the bronchioles that reduces gas delivery; emphysema—a destruction of the alveoli that reduces oxygen and carbon dioxide exchange; and lung cancer—a destruction of lung tissue. Since the respiratory and circulatory systems are so closely linked, any burden placed on the respiratory system will potentially affect the circulatory system. *Cardiac implications* are sometimes very noticeable, particularly during severe air pollution episodes. Carbon monoxide can kill or can have serious effects on chronic heart and lung diseases.

The lungs remove unreactive gaseous pollutants; however, the reactive gases may remain long enough to cause irritation of the lung tissues. Irritation causes weakening, which makes the damaged tissues more susceptible to pollutants that enter the lungs, and thereby contributes to the lung diseases previously mentioned. Pollutant gases may pass ino the circulatory system and be excreted by the lungs, kidneys, or intestines. Deposition somewhere in the body system may occur and cause further damage unless neutralized by one of the body's protective mechanisms. The protective hair-like cilia in the upper intestinal tract may be damaged by sulphur dioxide and nitrogen dioxide air pollutants.

Because of the complexity of the air pollution mix in the atmosphere, determining the results of interaction of pollutants related to bodily injury is very difficult. The effect of pollutants in the eyes and lungs is the most easily detectable. Knowledge of the functions of these target organ systems will aid in understanding the physiological effects of air pollutants on man.

PHYSIOLOGICAL EFFECTS ON MAN

The word "physiological" pertains to the functioning of the body organs. This section deals with specific air pollutants and how they disrupt healthy existence. Some of these potentially harmful contaminants have not been studied in depth; however, preliminary air pollution surveys have been made and a literature review has been documented by the U.S. Public Health Service (USPHS). Much of the following material has been extracted from USPHS Publications APTD 69–25 to 69–49.

biological pollutants

Aeroallergens are airborne materials that elicit a hypersensitivity response in susceptible individuals. The two major responses are hay fever and bronchial asthma. Less common responses are acute skin rash characterized by hives and eczema. Statisticians estimate that 10 to 15 million people in the United States suffer from seasonal hay fever and 5 to 10 percent of untreated hay-fever patients develop asthma. Trees are an important source of pollen in late winter and early spring, whereas grasses may be the primary causative agent during the summer months. Ragweed pollen causes more than 90 percent of hay fever in the United States, with fall being the major seasonal period. Table 3–1 shows the common aeroallergens and the source of each.

Biological aerosols are suspensions of microorganisms in the air, such as bacteria, fungi, and viruses, which may cause diseases in humans, animals, and plants, and degradation of inanimate materials.

These airborne microorganisms come from the soil, from plants, and from water. Also, animals set microorganisms free through shedding, excretion, and perspiration. Most microorganisms in the air are saprophytic (live on dead organic matter) and generally not pathogenic (capable of causing diseases); however, most microorganisms coming from living organisms are pathogens and are found close by the host. Because microorganisms do not survive very long as aerosols, airborne transmission of human diseases is limited to indoor (such as hospitals, homes, offices, and schoolrooms) or crowded outdoor spaces where the host comes in close proximity to the receptor. The microbic contamination of air by sewage treatment plants and animal-rendering plants has also been recorded.

Airborne bacterial infections of humans are listed in Table 3–2.

Airborne fungal infections of humans are listed in Table 3–3.

Airborne transmissions of viral diseases, *other than respiratory diseases,* are mumps, german measles (rubella), red measles (rubeola), small pox (variola), chicken pox (varicella), and shingles (herpes zoster). Airborne viral *respiratory diseases* of humans are listed in Table 3–4.

TABLE 3-1   Common Aeroallergens

Aeroallergens	Source
Pollens	Wind-pollinated plants, grasses, weeds, and trees
Molds	Usually saprophytic, prevalence depending upon humidity
Danders	Feathers of chickens, geese, ducks; hair of cats, dogs, horses, sheep, cattle, laboratory animals, and humans
House dust	A composite of all dusts found about the home
Miscellaneous vegetable fibers and dusts	Cotton, kapok, flax, hemp, jute, straw, castor beans, coffee beans, oris root, rye, wheat
Cosmetics	Wave set lotions, talcs, perfumes, hair tonics
Insecticides	Insecticides containing pyrethrum as a common ingredient
Paints, varnishes, and glues	Linseed oil and organic solvents

Extracted from Preliminary Air Pollution Survey of Aeroallergens, APTD 69-23.

non-biological pollutants

Reports indicate that high levels of air pollution correlate with high incidences of influenza and other respiratory illnesses. This tends to point out that the two types of pollution, biological and nonbiological, may produce synergistic (related to cooperative action of two agents such that the total effect is greater than the sum of the two effects taken independently) or potentiating (augmenting or additional) effects. The nonbiological pollutants discussed next are summarized in Table 3–5, indicating source and effects.

Aldehydes are products of incomplete combustion of hydrocarbons and other organic materials. Formaldehyde and acrolein are the two major aldehyde air pollutants.

TABLE 3-2 Common Airborne Bacterial Infections of Humans

Disease	Causative Agent	Symptoms and Remarks
Pulmonary tuberculosis	Mycobacterium tuberculosis	Lesions caused by nodules or tubercles are found in the lungs (or other parts of the body). In some cases calcification of the nodules takes place, and in others there is a coalescence of the necrotic tissue.
Pulmonary anthrax	Bacillus anthracis	Primarily a disease of animals but also occurs in man. This is the most dangerous, although not the most common, of the three forms of anthrax. It is characterized by many of the symptoms of pneumonia and often progresses into fatal septicemia.
Staphylococcal respiratory infection	Staphylococcus aureus	Can result in a gradual cavitating pneumonia or a fulminating hemorrhagic pneumonia.
Streptococcal respiratory infection	Streptococcus pyogenes	May develop into any of a variety of symptoms, including tonsillitis, sinusitis, otitis media, bronchopneumonia, pharyngitis, or septic sore throat, and becomes scarlet fever if the infecting strain produces erythrogenic toxin.
Meningococcal infection	Neisseria meningitidis	Probably becomes established initially in the nasopharynx but clinically develops into a cerebrospinal meningitis.
Pneumococcal pneumonia	Diplococcus pneumoniae	Clinically is nearly always lobar pneumonia. However, the infection may migrate through the nasal passages or be distributed via the vascular system to various parts of the body and give rise to localized foci of infection. Death is due to overwhelming interference with respiration or to general systematic toxemia.

TABLE 3-2   (Cont'd)

Disease	Causative Agent	Symptoms and Remarks
Pneumonic plague	Pasteurella pestis	Although ordinarily spread by the bite of fleas, it can occur secondary to glandular plague and give rise to a primary pulmonary form transmitted from man to man; usually fatal.
Whooping cough	Bordetella pertussis	Usually a childhood disease which begins with a catarrhal stage of a mild cough that progresses in severity to a paroxysmal stage characterized by rapid consecutive coughs and the deep inspiratory whoop. In the convalescent stage, the number and frequency of paroxysms gradually decrease.
Diphtheria	Corynebacterium diphtheriae	A childhood disease, usually a local infection of the mucous surfaces. The pharynx is most commonly affected, but infection of the larynx, or membranous croup, and nasal diphtheria are not infrequently observed. Primary infection of the lungs and other parts of the body has been reported.
Klebsiella pulmonary infection	Klebsiella pneumoniae	Produces necrotic lesions of the lung parenchyma and usually is fatal if not treated.
Staphylococcal wound infection	Staphylococcus aureus	Those surgical wounds which become infected by bacteria settling from air in surgery room. These organisms may be derived from the surgical team or may be carried into the operating room by air currents.

Extracted from Preliminary Air Pollution Survey of Biological Aerosols, APTD 69-30.

Ammonia is harmful to the body and can be fatal when released into the atmosphere in high concentrations. It is believed that zinc ammonium sulfate aerosols were in part responsible for the irritant effects of the air disaster at Donora in 1948.

TABLE 3-3   Common Airborne Fungal Infections of Humans

Disease	Causative Agent	Symptoms and Remarks
Blastomycosis	Blastomyces dermatitidis	A chronic granulomatous mycosis clinically resembling tuberculosis with coughing, pain in the chest, and weakness.
Coccidioido-mycosis	Coccidioides immitis	Varies in severity in recognized primary cases from that of a common cold to cases resembling influenza. Many cases are symptomless. The secondary or progressive coccidioido-mycosis results in cutaneous, subcutaneous, visceral, and osseous lesions with a high fatality rate.
Cryptococcosis	Cryptococcus neoformans	More commonly is a generalized infection, but can also be a primary (or secondary) lung infection. It may spread from the lungs as well.
Histoplasmosis	Histoplasma capsulatum	A systemic mycosis of varying severity, with the primary lesion usually in the lungs. Clinical symptoms of the systemic form can resemble many other diseases (anemia, leukopenia, Hodgkin's disease, etc.).
Nocardiosis	Nocardia asteroides	A chronic disease resembling tuberculosis, often initiated in the lungs but sometimes progressing to a systemic infection.
Aspergillosis	Aspergillosis fumigatus	A chronic pulmonary mycosis similar to and sometimes mistaken for tuberculosis. The infection may be secondary, particularly to tuberculosis. Pulmonary infection results from inhalation of airborne spores.
Sporotrichosis	Sporotrichum schenckii	A nodular skin infection ultimately forming a necrotic ulcer. Transmission by inhalation of spores is rare.

Extracted from APTD 69-30.

TABLE 3-4 Viral and Related Agents Presently Recognized as the Cause of Human Respiratory Diseases

Group	No. Serotypes Causing Respiratory Illness	Serotype Name	Types of Clinical Syndromes Produced	Comments
Myxoviruses	2	Influenza A	Influenza, febrile pharyngitis or tonsillitis	Causes influenza in persons of all ages
		Influenza B	Common cold, croup, bronchitis, bronchiolitis	
	?	Influenza C	Pneumonia	
	1	Respiratory Syncytial (RS)	Bronchiolitis (infants), pneumonia, bronchitis, common cold, croup	Most common cause of bronchiolitis in children
	4	Parainfluenza	Croup (infants), bronchitis, common cold, pneumonia, bronchiolitis	Type 1 is the most important agent in the croup syndrome
Adenoviruses	8	1, 2, 3, 4, 5, 7, 4, 21	Bronchitis, common cold, pneumonia, bronchiolitis, febrile sore throat	
Picornaviruses	7	Coxsackie A (2, 3, 5, 6, 8, 10, 21)	Febrile sore throat, common cold	

				Most frequently isolated viruses in adults with upper respiratory infections
	3	Coxsackie B (2, 3, 5)	Febrile sore throat, common cold, pleurodynia	
	60+	Rhinoviruses	Common cold, bronchitis, pneumonia	
	2	ECHO (11, 20)	Febrile sore throat, common cold, croup	
Reoviruses (classification uncertain)	3	Reovirus (ECHO-11)	Minor respiratory symptoms and diarrhea (children)	
Herpesviruses	3	Herpes	Pharyngitis (adults)	
	1	Varicella	Pneumonia	
Chlamydozoaceae*	?	Psittacosis	Psittacosis, pneumonia	
Mycoplasmataceae	1	Mycoplasma	Pneumonia (Eaton agent), bronchitis, bronchiolitis, minor upper respiratory illness	
Rickettsiae		Coxiella burnetii (Q fever)	Pneumonia	

Extracted from APTD 69-30.

* Not a true virus; nucleic acid core contains both RNA and DNA.

TABLE 3-5   Non-Biological Air Pollutant Effects on Man

Pollutant	Source	Effects
Aldehydes (formaldehyde and acrolein)	Auto exhaust, waste incineration, fuel combustion; photochemical reactions	Eye, skin and respiratory irritation, odors
Ammonia	Chemical industries, coke ovens, refineries, stock yards and fuel incineration	Corrosive to mucous membranes, damage to eye and respiratory tract
Arsenic	Metal smelters, arsenical pesticides and herbicides	Inhaled, ingested or absorbed through skin; causes dermatitis, mild bronchitis and nasal irritation, suspected carcinogen
Asbestos	Asbestos factory or mine, construction sites	Pulmonary fibrosis, pleural calcification, lung cancer
Barium	Industries mining, refining or producing barium and barium-based chemicals; smoke suppressant additives in diesel fuels	Affects heart muscles, gastrointestinal tract and central nervous system and respiratory tract
Beryllium	Industrial usage, production of fluorescent lamps; rocket motor fuels	Pulmonary damage and damage to skin and mucous membranes from handling soluble salts of beryllium
Boron	Industry producing boron, petroleum fuel additive, present in coal	Toxic through ingestion or inhalation, as dust causes irritation and inflammation. Boron hydrides can cause damage to central nervous system and death.
Cadmium	Metal industries engaged in extraction, refining, machining, electroplating and welding of cadmium materials. By-product of refining lead, zinc and copper. Pesticides, fertilizers, cadmium-nickel batteries, reactor poison in nuclear fission plants, used in production of tetraethyl lead gasoline.	Chronic and acute poisoning. Inhalation of fumes and vapors causes damage to kidneys, emphysema, bronchitis, cancer, gastric and intestinal disorders, diseases of heart, liver and brain.
Chlorine	Process industries using chlorine, accidental leakage during storage and transportation	Irritates eye, nose and throat. Large doses damage lungs producing edema, pneumonitis, emphysema and bronchitis.
Chromium	Metallurgical and chemical industries. Products em-	Irritative, corrosive and toxic to body tissues. Be-

TABLE 3-5   (Cont'd)

Pollutant	Source	Effects
	ploying chromate compounds, cement, asbestos	lieved to exert carcinogenic action. Develops dermatitis and ulcers of the skin and perforation of nasal septum.
Ethylene	Motor vehicle emissions, chemical industries, incineration of agricultural wastes, emitted by growing plants	Eye irritation as a result of photochemical reaction with nitrogen oxides and ozone
Hydrochloric acid	By-product from chlorination of organic compounds, burning of coal, burning of chlorinated plastics and paper, combustion of gasoline containing ethylene chloride	Coughing and choking from inhalation; inflammation and ulceration of upper respiratory tract, clouding of the cornea
Hydrogen sulfide	Biological decay of protein material in stagnant water. Kraft paper mills, industrial waste disposal ponds, sewage treatment plants, refineries, coke ovens.	Headaches, conjunctivitis, sleeplessness, pain in the eyes. Odorous. In high concentration can lead to blockage of O_2 transfer, act as a cell and enzyme poison and damage nerve tissues.
Iron	Iron and steel plants, fly ash from combustion of coal and fuel oil, municipal waste incineration, use of welding rods	Benign pneumoconiosis and siderosis (iron pigmentation of the lungs). Iron oxides act as a vehicle for transporting carcinogens and sulfur dioxide deep into the lungs.
Lead	Automobile emissions when using leaded gasoline, lead smelters, combustion of coal and fuel oil, lead-arsenate pesticides	Absorbed through gastrointestinal and respiratory tract and deposited in mucous membrane of nose, throat and in the lungs
Manganese	Blast furnaces producing ferromanganese compounds, organic manganese fuel additives, use of welding rods, incineration of manganese-containing products	Manganese poisoning of central nervous system, manganic pneumonia absorbed by inhalation, ingestion or through the skin
Mercury	Mining and refining of mercury, use of mercury in laboratories. Pesticides containing mercury.	Vapors inhaled may cause intoxication or protoplasmic poisoning.

TABLE 3-5   (Cont'd)

Pollutant	Source	Effects
Nickel	Metallurgical plants using nickel, engines burning fuels containing nickel additives, burning coal and oil, nickel plating facilities, incineration of nickel products	May cause cancer of the lung, cancer of the sinus, other respiratory disorders, dermatitis
Phosphorus	Plants producing phosphate fertilizer, phosphoric acid, phosphorus pentoxide. Emission from vehicles and airplanes using phosphorus as corrosion inhibitors in fuel.	Skin irritation to systemic poisoning. High concentrations affect nervous system.
Radioactive substances	Direct contamination by radioactive gases or suspended dust from natural or artificial sources. Indirect contamination when radionuclides are ingested from food chain as a result of contaminated ground, water, plants or animals.	Somatic effects such as leukemia and other types of cancer, cataracts and reduction in life expectancy. Genetic effects include mutations in human gametes that show up only in future generations.
Selenium	Combustion of industrial and residual fuels, refinery waste gases and fumes, incineration of wastes including paper products	Irritation of eye, nose, throat, respiratory tract and gastrointestinal tract. Chronic intoxication from industrial exposure has possible long term effects on kidney, liver and lungs.
Vanadium	Vanadium refining industries, alloy industries, power plants and utilities using vanadium-rich residual oils	Physiological effects of varying severity on the gastrointestinal and respiratory tracts. Inhibition of cholesterol synthesis. Chronic exposure related to heart disease and cancer.
Zinc	Zinc refineries, manufactures of brass, zinc galvanizing processes	Fumes have corrosive effect on the skin and irritate and damage mucous membranes.

Arsenic and its compounds are known to be toxic to humans, animals, and plants. Poisoning from arsenical dusts in the normal atmosphere is unlikely; however, when proper control devices are not maintained on arsenic metal smelters, poisoning episodes can occur. Arsenical herbicides are used to eliminate aquatic plants; aquatic animals have a higher tolerance for arsenic and are not killed. If not properly controlled, arsenical herbicides can become an air pollution problem. Arsenic is also used as a desiccant (drying agent) to remove cotton leaves prior to machine picking. As a result, the ginning of cotton and the burning of cotton trash after the use of arsenic causes air pollution problems. The burning of coal also releases a small amount of arsenic into the air.

Asbestos is the general name given to a variety of useful fibrous minerals. The major asbestos minerals are chrysotile, crocidolite, amosite, and anthophyllite. Inhalation of asbestos dust is an industrial hazard as well as a nonoccupational environmental hazard for persons living near an asbestos factory or mine. According to one researcher, lung cancer deaths among asbestos workers who smoked cigarettes were eight times higher than for nonsmokers exposed to asbestos. The reason is that asbestos fibers irritate membrane; continuous contact leads to tumors, which are susceptible to the tars and nicotine inhaled from cigarette smoking.

Barium is a soft, silvery metallic element found in nature only in combination with other elements. Insoluble compounds such as barium sulfate are generally nontoxic; however, the soluble compounds are highly toxic when ingested, and produce a strong stimulating effect on all muscles of the body. More research is needed to evaluate barium as a possible air pollutant, especially in its role as a smoke suppressant in diesel engine fuel.

Beryllium and its compounds are of concern because of their highly toxic effect on the health of humans. Beryllium is considered among the most toxic and most hazardous of the nonradioactive substances being used in industry. Expanding industrial use of beryllium suggests that more research is needed to properly evaluate beryllium as an air pollutant.

Boron and its compounds are considered moderately to highly toxic to man when ingested or inhaled. More research is needed regarding boron as an environmental contaminant.

Cadmium and cadmium compounds are considered toxic to humans; however, the exact nature of the biological action of cadmium is not fully understood. Recent studies indicating cadmium may be a cause of heart disease and cancer have given researchers concern about low concentrations of cadmium in the atmosphere. Henry A. Schroeder, Director of the Trace Element Laboratory at Dartmouth Medical School, advises that the evidence is clear that cadmium is a major factor contributing to hypertension. Most atmospheric cadmium evolves from zinc refineries.

Chlorine is a dense, greenish-yellow gas with a distinctive, irritating odor. Chlorine gas is a very strong oxidizing agent, capable of reacting with both organic and inorganic materials. This makes it dangerous to humans, animals, plants, and numerous materials. In the presence of moisture it reacts to form hypochlorite and hydrochloric acid.

Chromium is thought to be relatively nontoxic in metal form, but when trivalent and hexavalent chromium compounds are found in the air, they are toxic to humans. Chromium trioxode (CrO_3) is perhaps the most important hexavalent chromium compound in the air, because when added to water it produces chromic acid (H_2CrO_4), which is very corrosive.

Ethylene is a colorless hydrocarbonic gas of the olefin series which, although not toxic to humans, presents a considerable air pollution problem. It is the most abundant of the hydrocarbons in the lower atmosphere. Ethylene contributes to photochemical air pollution, and when it combines with nitrogen oxides and ozone, the reactions cause eye irritations.

Hydrochloric acid is an aqueous solution of hydrogen chloride, which is a hygroscopic, colorless gas with a strong, pungent, and irritating odor. There are no known chronic or acute systemic effects of hydrochloric acid; however, the strong dehydrating properties of hydrogen chloride (HCl) can result in serious injuries to man.

Hydrogen sulfide is a colorless gas that has an obnoxious rotten-egg odor even at low concentrations.

Iron and its compounds present as pollutants in the atmosphere can cause deleterious effects to humans.

Lead is a known poison. Alkyl lead compounds, used primarily as automotive fuel additives, are readily absorbed by skin and mucous membranes and preferentially distributed to the lipid (fats and related compounds) phases of the body, including the brain. About 50 percent of lead inhaled is retained, and when the quantity of lead reaches a high enough level, it interferes with development of red blood cells and the production of hemoglobin. Atmospheric lead pollution is an important additive to the overall lead input to man's environment and must be considered to have potential health effects.

Manganese or its compounds when inhaled may produce chronic manganese poisoning (a disease of the central nervous system) or manganic pneumonia, which has a mortality rate four times that for lobar pneumonia. It may also act as a catalyst in the oxidation of other air pollutants, producing even more undesirable pollutants.

Mercury is a liquid metal at normal temperatures that emits vapors into the atmosphere, causing pollution. Inhalation of mercury can be even more dangerous than entry through the digestive tract.

Nickel and its compounds, taking the form of dust and vapors, can be serious when inhaled or absorbed through the skin.

Phosphorus and its compounds have varying effects, depending upon their chemistry and concentration. Tests have been made on the effects of yellow (white) phosphorus, but acute or chronic effects have been studied only with respect to organophosphorus pesticides.

Radioactive substances can cause injury to man by exposure from a distant source, direct contact with the skin, or entrance into the body when radionuclides (e.g., strontium 90, iodine 131, cesium 137) are ingested from food contaminated by these radioactive substances.

Selenium poisoning of humans from ingestion of foods containing toxic amounts of selenium is a problem of great concern in the United States. Selenium may enter these foods from air pollution resulting from fuel combustion, fuel refinery waste gases, and incineration of paper products containing selenium.

Vanadium is moderately toxic to humans, especially in its pentavalent form.

Zinc and its compounds are generally considered to be nontoxic to humans except in high concentrations; however, inhalation of zinc oxide fumes produces "metal-fume," which is a nonfatal but self-limiting illness. Accidental poisoning can result from ingestion of acidic foods prepared in zinc-galvanized containers. Since zinc is commonly associated with other metals in the air, it is difficult to assess its value as a separate air pollutant.

odorous pollutants

In addition to the preceding effects, various combinations of organic and inorganic substances are referred to as odorous compounds since they may produce pleasant or unpleasant odors. Malodors (bad odors) are one of the first manifestations of air pollution and frequently arouse extreme emotional reactions. These odors cause mental and physiological effects such as nausea, headache, loss of sleep, loss of appetite, impaired breathing, and, in some cases, allergic reactions. Hydrogen sulfide, aldehydes, chlorine, hydrochloric acid, ammonia, and ethylene were odorous substances covered in previous paragraphs. The most frequent sources of odors are listed in Table 3–6.

organic carcinogens

Organic carcinogens fall into three main categories: polynuclear aromatic hydrocarbons, polynuclear heterocyclic compounds and oxygenated compounds, and alkylating agents. There is an apparent association between lung cancer and increased tumor activity and organic carcinogens. The major emission sources are heat generation, refuse burners, motor vehicle exhausts, and industrial processes.

chemical pesticides

Chemical pesticides or economic poisons include a spectrum of chemicals used to control or destroy pests that cause economic losses or adverse human health effects. These can be grouped as insecticides, (weed and brush killers, defoliants, and desiccants), fungicides, rodenticides, ascaricides, nematocides, molluscacides, algicides, repellants, attractants, and plant growth regulators, according to their specific use. These pesticides are employed in agriculture, forestry, food storage, urban sanitation, and in the home. Although intended to be toxic for only certain forms of life, these pest killers when used improperly can cause adverse effects on humans. They can enter the body by ingestion, inhalation, and absorption through the skin.

The chlorinated hydrocarbons and organophosphorus insecticides are the greatest health hazards because of inherent toxicity or persist-

TABLE 3-6   Most Frequently Reported Odor Sources

Source of Odor	Number Reported
Animal Odors	
Meat packing and rendering plants	12
Fish oil odors from manufacturing plants	5
Poultry ranches and processing	4
Odors from combustion processes	
Gasoline and diesel engine exhaust	10
Coke-oven and coal-gas odors (steel mills)	8
Poorly adjusted heating systems	3
Odors from food processing	
Coffee roasting plants	8
Restaurants	4
Bakeries	3
Paint and related industries	
Manufacturing of paint, lacquer and varnish	8
Paint spraying	4
Commercial solvents	3
General chemical odors	
Hydrogen sulfide	7
Sulfur dioxide	4
Ammonia	3
General industrial odors	
Burning rubber from smelting and debonding	5
Odors from dry-cleaning shops	5
Fertilizer plants	4
Asphalt odors (roofing and street paving)	4
Asphalt odors (manufacturing)	3
Plastic manufacturing	3
Foundry odors	
Coke-oven odors	4
Heat treating, oil quenching and pickling	3
Smelting	2
Odors from combustion of waste	
Home incinerators and backyard trash fires	4
City incinerators burning garbage	3
Open-dump fires	2

TABLE 3-6   (Cont'd)

Source of Odor	Number Reported
Refinery odors	
Mercaptans	3
Crude oil and gasoline	3
Sulfur	1
Odors from decomposition of waste	
Putrefaction and oxidation (organic acids*)	3
Organic nitrogen compounds (decomposition of protein*)	2
Decomposition of lignite (plant cells)	1
Sewage odors	
City sewers carrying industrial waste	3
Sewage treatment plants	2

Extracted from Preliminary Air Pollution Survey of Odorous Compounds, APTD 69-42.
* Probably related to meat processing plants.

ence of residues. Cases of poisoning, both accidental and occupational, have been reported for practically every known insecticide. Symptoms of poisoning in mild cases are characterized by headaches, dizziness, stomach disturbances, and high irritability. In most serious cases, muscular tremors appear with jerky movements leading to convulsions and death. In most cases it is difficult to determine if exposure was predominantly respiratory or dermal (through the skin).

The economic and social benefits gained by the use of pesticides are great; yet ill effects causing death to animals may be a precursor of danger to humans.

The previous coverage dealt with physiological effects of specific air pollutants. The following coverage deals with combinations of these elements in the form of inorganic gases, organic gases, photochemical oxidants, and particulates of greatest concern. Much emphasis has been placed on these pollutants, and pollution control agencies have established ambient air quality standards related to most of these pollutants.

inorganic gases

Carbon monoxide (CO), produced primarily by the incomplete combustion of fuels, is an asphyxiating gas in enclosed places, such as tunnels, and highways with high-density traffic. Carbon monoxide in

TABLE 3-7   Effects of Carbon Monoxide

Environmental Conditions	Effect	Comment
35 mg/m^3 (30 ppm) for up to 12 hours	Equilibrium value of 5% blood COHb is reached in 8 to 12 hrs. 80% of this equilibrium value (4% COHb) is reached within 4 hrs.	Experimental exposure of nonsmokers. Theoretical calculations suggest exposure to 23 (20 ppm) and 12 mg/m^3 (10 ppm) would result in COHb levels of about 3.7 and 2%, respectively, if exposure was continuous for 8 or more hours.
58 mg/m^3 (50 ppm) for 90 minutes	Impairment of time-interval discrimination in nonsmokers.	Blood COHb levels not available, but anticipated to be about 2.5%. Similar blood COHb levels expected from exposure to 10 to 17 mg/m^3 (10 to 15 ppm) for 8 or more hours.
115 mg/m^3 (100 ppm) intermittently through a facial mask	Impairment in performance of some psychomotor tests at a COHb level of 5%.	Similar results may have been observed at lower COHb levels, but blood measurements were not accurate.
High concentrations of CO were administered for 30 to 120 secs. and then 10 min. was allowed for washout of alveolar CO before blood COHb was measured.	Exposure sufficient to produce blood COHb levels above 5% has been shown to place a physiologic stress on patients with heart disease.	Data rely on COHb levels produced rapidly after short exposure to high levels of CO; this is not necessarily comparable to exposure over a longer time period or under equilibrium conditions.

Extracted from Air Quality Criteria for Carbon Monoxide, AP-62.

the ambient atmosphere may cause physical problems not necessarily leading to death, such as headaches, loss of visual acuity, loss of ability to accurately estimate time intervals, decreased muscular coordination, and loss of oxygen from the blood. The affinity of hemoglobin to absorb CO is over 200 times that for O_2; therefore, CO will be absorbed in preference to O_2, causing a reduction in the O_2 carrying capacity of blood. A CO atmospheric level of 35 mg/m³ (30 ppm) has been equated to the loss of 1 pint of blood. In other words, 1 pint of blood would be tied up with the CO and not be available to carry O_2. This can be serious for persons with anemia or cardiac or respiratory disease. Eight hours of exposure at 92 mg/m³ (80 ppm) CO will render 15 percent of the hemoglobin useless for carrying O_2; 20 percent of humans will have respiratory difficulties, 15 percent will be dizzy, 30 percent will have headaches, and nearly everyone will need a light 60 percent brighter before he can see it.

Present national ambient air quality standards are established at 10 mg/m³ for a maximum 8-hour(h) concentration or limited to 40 mg/m³ CO over a 1-h period. Table 3–7 lists effects of CO.

Sulfur oxides (SO_2, SO_3), produced mainly by the burning of coal and fuel oil, and the acids resulting from combination of these oxides with water (H_2SO_3 and H_2SO_4), and the salts derived from these acids when combined with other elements are well-known atmospheric pollutants. The dominant pollutant in this group is sulfur dioxide (SO_2) and is stressed in this coverage of effects on man. Increased mortality may occur at 24-h averages of 1500 μg/m³ of SO_2. Increased death rates may occur at a 24-h mean of 500 to 715 μg/m³ of SO_2. Adverse health effects have been noted when annual mean levels of SO_2 exceed 115 μg/m³ or when 24-h average levels exceed 300 μg/m³.

Analysis of numerous epidemiological studies in London and New York City regarding effects of SO_2 clearly indicates a synergistic effect between SO_2 and particulates and the varying severity of health effects. Public Health Service Publication No. AP-50, *Air Quality Criteria Pamphlet* (Sulfur Oxides), indicates that a weight per volume measurement of 715 μg/m³ of SO_2 over a 24-h period combined with a smoke concentration of 750 μg/m³ in the same 24-h period causes greater health damage than either of these two conditions when they occur separately. Also, 1,500 μg/m³ of SO_2 over a 24-h period combined with particulates measured as a soiling index of 6 COH (coefficient of haze) causes increased mortality over either condition when occurring separately. (A COH is a measure of light absorption by particulates defined as that quantity of light-scattering solids producing an optical density of 0.1.)

Intense irritation, reduction of visibility, contribution to respiratory diseases, and cardiac ailments have also been recorded from epidemiological studies pertaining to sulfur oxides.

Present national ambient air quality standards are established at 80 $\mu g/m^3$ annual arithmetic mean, or 365 $\mu g/m^3$ maximum 24-h concentration not to be exceeded more than once per year.

Nitrogen oxides (NO and NO_2) are formed when combustion at high temperatures results in the fixation of nitrogen and oxygen. The summation of the two is NO_x. NO_x related to photochemical smog is covered under *oxidants*. NO is not an irritant and is not considered to have adverse health effects at normal atmospheric concentrations. The greatest significance of NO is related to its tendency to undergo oxidation to NO_2.

Affinity of hemoglobin for absorbing NO_2 is 300,000 times that for O_2. This affinity drastically reduces the O_2 carrying capacity of the blood. High levels can kill; short-term exposures of NO_2 for less than 24–h continuously can have several concentration-dependent effects: the olfactory threshold value is about 225 $\mu g/m^3$ (0.12 ppm); exposure to 9.4 mg/m^3 (5 ppm) for 10 min has produced transient increase in airway resistance, and occupational exposure to 162.2 mg/m^3 (90 ppm) for 30 min has produced pulmonary edema (watery swelling) 18-h later, accompanied by an observed vital capacity that was 50 percent of the value predicted for normal function. NO_2 is the basic pulmonary irritant. Table 3–8 lists a summary of representative NO_2 effects.

Long-term exposure to NO_2 at concentrations between 117 and 205 $\mu g/m^3$ and mean suspended nitrate level at 3.8 $\mu g/m^3$ results in acute respiratory disease. Acute bronchitis among infants and school-children results with slightly lower concentrations, due to more tender and susceptible tissues than in adults.

It has been stated that 95 percent of nitrogen oxides inhaled remain in the body, where they *can also produce mutations in cells.* Nitrogen oxides cause lung tissues to become leathery and brittle and may cause lung cancer and emphysema. The American Medical Association notes that increased death rates for emphysema are more closely correlated with the growth rate of vehicles than with cigarette smoking. National ambient air quality standards have been set at 100 $\mu g/m^3$ annual arithmetic mean.

inorganic photochemical oxidants

Inorganic photochemical oxidants (ozone and nitrogen dioxide) are produced by the radiant energy of the sun changing primary pollutants into secondary pollutants by various reactions.

TABLE 3-8   Summary of Representative NO$_2$ Effects

Effect	NO$_2$ Concentration ppm	NO$_2$ Concentration $\mu g/m^3$	Duration	Comment
Lowest level associated with reference oxidant production of 200 $\mu g/m^3$ (0.1 ppm)	0.04	80	3 hr. (6 to 9 A.M.)	
Increased incidence of acute respiratory disease in families	0.062 to 0.109	117 to 205	2 to 3 yr.	Chattanooga study—6 mo. mean concentration range
Increased incidence of acute bronchitis in infants and school children	0.063 to 0.083	118 to 156	2 to 3 yr.	Chattanooga study—6 mo. mean concentration range
Human olfactory threshold	0.12	225		Immediate perception
Man—increase in airway resistance	5	9400	10 minutes	Transient

Extracted from Air Quality Criteria for Nitrogen Oxides, AP-84.

Ozone (O_3) at times makes up 99 percent of the oxidants that are produced photochemically in the air. Concentrations of ozone exceeding 196 $\mu g/m^3$ (0.1 ppm) will cause eye irritation; the threshold for both nasal and throat irritation is set higher at 0.3 ppm over an 8-h period. Some states permit 0.15 ppm for 1-h periods. When the level reaches 0.3 to 1.00 ppm over a 15-min to 2-h period, choking, coughing, and severe fatigue will ensue. Levels of 1.5 to 2.0 ppm over a 2-h exposure will cause severe chest pains, coughing, headache, loss of coordination, and difficulty in expression and articulation. Exposures to 9.0 ppm produce severe illness. See Table 3–9 for effects of ozone.

Nitrogen oxides are considered under inorganic photochemical oxidants because the formation of O_3 is dependent upon NO_2 to a certain degree when NO_2 is broken down by ultraviolet light energy to NO and O; the subsequent reaction of the O atom is to combine with the O_2 molecule of the atmosphere to form O_3. Collisions with O_3 and other molecules transfer excess energy, leading to formation of a stable O_3 molecule. National ambient air quality standards for photochemical oxidants have been set at 160 $\mu g/m^3$ for a maximum 1-h concentration not to be exceeded more than once per year.

organic gases

Total oxidants in the air may be defined as those compounds that will oxidize a reference material which is not capable of being oxidized by atmospheric oxygen. However, since ambient air contains a mixture of oxidizing and reducing agents (O_3, NO_2, PAN, SO_2, H_2S, aldehydes, unsaturated hydrocarbons, and others), and these reducing agents and oxidizing agents have an opposite effect on the reference material, the result obtained is a "net" oxidant rather than a total oxidant value. Oxidant concentrations of 250 $\mu g/m^3$ (0.13 ppm) cause an increase in asthmatic attacks over long-term exposures, and short-term exposures of 200 $\mu g/m^3$ (0.1 ppm) cause eye irritation. Table 3–10 lists effects associated with oxidant concentrations in photochemical smog.

Gaseous hydrocarbons alone have not demonstrated direct adverse effects on human health; however, they play a very important part in the formation of photochemical oxidants that do produce adverse effects. Gaseous hydrocarbons may be broken down into these general categories: the *aliphatic* (including olefins and paraffins) and the *alicyclic* hydrocarbons, which are generally biologically active although biochemically inert, and the *aromatic* hydrocarbons, which are both biologically and biochemically active and their vapors the most

TABLE 3-9   Effects of Ozone

Effect	Exposure		Duration	Comment
	ppm	$\mu g/m^3$		
Odor detection	0.02	40	5 minutes	Odor detected in 9 of 10 subjects
Increased susceptibility of laboratory animals to bacterial infection	0.08 to 1.30	160 to 2,550	3 hours	Demonstrated in mice at 160 $\mu g/m^3$ and in mice at 2550 $\mu g/m^3$
Respiratory irritation (nose and throat), chest constriction	0.30	590	Continuous during working hours	Occupational exposure of welder, other pollutants probably present
Changes in pulmonary function; diminished $FEV_{1.0}$ after 8 weeks	0.50	980	3 hours/day, 6 days/week, for 12 weeks	Change returns to normal 6 weeks after exposure. No changes observed at 390 $\mu g/m^3$ (0.2 ppm).
Small decrements in VC, FRC, and DL_{CO} in, respectively, 3, 2, and 1 out of 7 subjects	0.20 to 0.30	390 to 590	Continuous during working hours	Occupational exposure. All 7 subjects smoked. Normal values for VC, FRC, and DL_{CO} based on predicted value.

Effect	Concentration (ppm)	Concentration ($\mu g/m^3$)	Duration	Comments
Impaired diffusion capacity (DL_{CO})	0.60 to 0.80	1,180 to 1,570	2 hours	Experimental exposure of 11 subjects
Increased airway resistance	0.10 to 1.00	200 to 1,960	1 hour	Significant increase in 2 of 4 subjects at 200 $\mu g/m^3$ (0.1 ppm) and 4 of 4 subjects at 1960 $\mu g/m^3$ (1.0 ppm)
Reduced VC, severe cough, inability to concentrate	2.00	3,900	2 hours	High temperatures. One subject.
Acute pulmonary edema	9.00	7,600	Unknown	Refers to peak concentration of occupational exposure. Most of exposure was to lower level.

Extracted from Air Quality Criteria for Photochemical Oxidants, AP-63.

TABLE 3-10   Effects Associated with Oxidant Concentrations in Photochemical Smog

Effect	Exposure		Duration	Comment
	ppm	$\mu g/m^3$		
Eye irritation	Exceeding		Peak values	Result of panel response
	0.1	200		
Aggravation of respiratory diseases—asthma	0.25	490	Maximum daily value	Patients exposed to ambient air. Value refers to oxidant level at which number of attacks increased
				Such a peak value would be expected to be associated with a maximum hourly average concentration as low as $300\mu g/m^3$ (0.15 ppm)
Impaired performance of student athletes	Exceeding		1 hour	Exposure for 1 hour immediately prior to race
	0.07	130		

Extracted from Air Quality Criteria for Photochemical Oxidants, AP-63.

irritating to the mucous membranes. All the hydrocarbons can by photochemical reaction increase production of smog effects.

When NO_2 undergoes photolysis (chemical decomposition by the action of radiant energy), some of the free oxygen atoms released combine with hydrocarbons to form hydrocarbon free radicals. Combination of the acyl radical with NO_2 produces peroxyacetyl nitrate (PAN), a major eye irritant. The oxygenated hydrocarbons of particular importance in photochemical smog are the aldehydes. Formaldehyde (HCHO) causes odor irritation with a threshold for odor as low as 70 $\mu g/m^3$ (0.06 ppm) and eye irritation with a threshold of 0.01 to 1.0 ppm. HCHO also causes bronchial irritation similar to the effects of SO_2. Acrolein (CH_2CHCHO), one of the compounds used in the production of tear gas, can be detected by both odor and eye irritation at concentrations as low as 600 $\mu g/m^3$ (0.25 ppm). Hydrocarbons may also convert SO_2 and SO_3 to H_2SO_4 mist that irritates noses and throats in concentrations as low as 15 ppm. National ambient air quality standards for hydrocarbons have been set at 160 $\mu g/m^3$ maximum 3-h concentration.

particulates

Particulate matter, in addition to being a soiling and nuisance factor, acts as a catalyst for other pollutants. Other contaminants absorb the particles and cause damage to the lungs and the respiratory tract. The aromatic hydrocarbons are often found in particulate matter. Major air pollution incidents of the past have been linked to SO_2–SO_3 combined with acid aerosols and particles of soot and fly ash. Excess deaths occur when particulate concentrations reach 750 $\mu g/m^3$ in a 24-h period and are accompanied by SO_2 concentrations of 715 $\mu g/m^3$. Many of the specific elements covered previously (i.e., zinc, vanadium, iron, lead, nickel, beryllium, boron, cadmium, asbestos, arsenic, aeroallergens) are of concern as air pollutant particulate matter. National ambient air quality standards for particulates are set at 75 $\mu g/m^3$ annual geometric mean or 260 $\mu g/m^3$ maximum 24-h concentration not to be exceeded more than once per year.

SPECIFIC DISEASES TO WHICH AIR POLLUTION CONTRIBUTES

chronic pulmonary diseases

Lung cancer, the destruction of lung tissue, has been increasing rapidly in recent years. Although many factors are probably involved, the high mortality rate in urban areas points toward air pollution as a major contributing factor. In addition to the organic carcinogens present in the atmosphere, many inorganics (i.e., arsenic, asbestos, cadmium, chromium, nickel, vanadium, and radioactive materials) are known or suspected to be carcinogenic in nature.

Chronic bronchitis, which affects the efficiency of air delivery by reducing the diameter of the bronchioles, has been researched extensively in Great Britain. There it is considered the second most common cause of death in men 40 to 55 years of age; incidence has been found to vary directly with the amount of air pollution. The ailment is characterized by a cough for 3 months of the year for 2 successive years.

Bronchial asthma is defined as an increased responsiveness of the trachea and the bronchi to various stimuli, and it is manifested by widespread narrowing of the airways. This condition is often aggravated by air pollution: the Donora episode, New Orleans asthma epidemics, and a special form of Tokyo–Yokohama asthma have borne this out.

Emphysema, the fastest growing cause of death in the United States, is described as a progressive breakdown of alveolar air sacs in the lungs brought on by chronic infection or irritation of the bronchial tubes, paralysis of the cilia, increased viscosity of mucous, injury from violent coughing, and prolonged retention of inert air pollutants. Emphysema progressively diminishes the ability of the lungs to transfer oxygen to the bloodstream and carbon dioxide from it. Abnormal enlargement of the air spaces distal to the terminal nonrespiratory bronchiole accompanied by destructive changes of the alveolar walls are contributing factors. Studies demonstrate emphysema patients improve when they are protected from air pollution. The fact that the incidence of emphysema is greater in our cities than in rural areas points to air pollution as a contributing factor.

The incidence of the *common cold* and *pneumonia* also appears to be connected with air pollution, according to many investigations.

REFERENCES

Hatch, T. F., et al., *Pulmonary Deposition and Retention of Inhaled Aerosols.* New York: Academic Press, Inc., 1964.

Millard, N. D., et al., *Human Anatomy and Physiology*, 4th ed. Philadelphia: W. B. Saunders Company, 1956.

Product Engineering for Design Engineers. New York: McGraw-Hill, Inc., Dec. 19, 1966.

Stern, A.C., *Air Pollution,* Chapter 10, Section VII. New York: Academic Press, Inc., 1962.

U.S. Department of Health, Education, and Welfare, *Air Quality Criteria Pamphlet,* PHS (NAPCA), Pub. Nos.
> AP-49, Particulates, 1969.
> AP-50, Sulfur Oxides, 1969.
> AP-62, Carbon Monoxide, 1970.
> AP-63, Photochemical Oxidants, 1970.
> AP-64, Hydrocarbons, 1970.
Washington, D.C.: U.S. Government Printing Office.

U.S. Department of Health, Education, and Welfare, *Preliminary Air Pollution Survey—A Literature Review,* PHS (NAPCA), Pub. Nos. APTD 69–23 through 69–49. Washington, D.C.: U.S. Government Printing Office, 1969.

U.S. Environmental Protection Agency, *Air Quality Criteria Pamphlet* (Nitrogen Oxides) Pub. No. AP-84, Washington, D.C.: U.S. Government Printing Office, 1971.

U.S. Environmental Protection Agency, *Environmental Lead and Public Health,* APCO, Pub. No. AP-90. Washington, D.C.: U.S. Government Printing Office, 1971.

U.S. Environmental Protection Agency, *Federal Register,* National Ambient AQ Standards, Vol. 36, No. 21, Part II, pp. 1502–1515. Washington, D.C.: U.S. Government Printing Office, Jan. 30, 1971.

RECOMMENDED FILMS

F-1528-X   Air Pollution and You (46-frame film strip)

MIS-678   Effects of Air Pollution (5 min)

MA-31   The Human Body Respiratory System (13 min)

MTS-773   Take a Deep Breath (25 min)
> Air Pollution Effects on Man's Respiratory System (set of 35-mm slides with guide)
> Available: Distribution Branch
> National Audio-visual Center (GSA)
> Washington, D.C. 20409

QUESTIONS

1/ Through what three types of investigation have we amplified our knowledge of the health effects of air pollution?

2/ Can it be said that air pollution can cause death?

3/ What advantage do epidemiological studies have over laboratory studies in determining air pollution effects on man?

4/ Why are data based on laboratory studies sometimes suspect in determining air pollution effects on man?

5/ What two target organ systems do scientists use to pinpoint relationships between air pollution and disease?

6/ What part of the respiratory system removes most particulates larger than 10 microns?

7/ What size particulates are most likely to be retained in the lungs?

8/ How many of these small particulates can eventually be removed?

9/ What three major types of lung damage may ensue when air pollution particles react with lung tissue? Describe each.

10/ What are some aeroallergens and what are their sources?

11/ What are the airborne microorganisms generally involved as biological aerosols that cause disease in man?

12/ What are some airborne fungal infections of humans?

13/ What are some airborne viral respiratory diseases of humans?

14/ What are some airborne bacterial infections of humans?

15/ What is a phytotoxicant?

16/ What is a major source of lead in the atmosphere?

17/ What are some odorous air pollutants?

18/ What are some airborne organic carcinogens?

19/ What are some suspected airborne inorganic carcinogens?

20/ What is a desiccant?

21/ What is a nematocide?

22/ What is the difference between radioactive somatic effects and genetic effects?

23/ What is meant by affinity of hemoglobin for a particular gas?

24/ Why is the concentration of CO usually indicated in milligrams per cubic meter whereas most other air pollutants are indicated in micrograms per cubic meter?

25/ What is the difference between average range and mean range?

26/ What is meant by mutations in cells?

27/ What is the most common photochemical oxidant in the air?

28/ What is the importance of gaseous hydrocarbons related to adverse effects on human health?

29/ In what form are aromatic hydrocarbons usually found in air pollution?

30/ What are four chronic pulmonary diseases discussed in their relationship to air pollution?

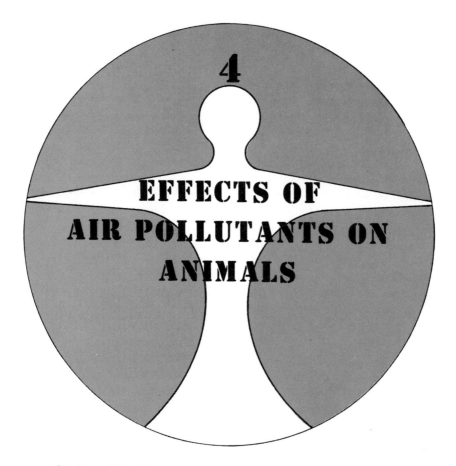

4

EFFECTS OF AIR POLLUTANTS ON ANIMALS

Student Objectives

—*To become aware of air pollution effects on animals.*
—*To understand the route of exposure in animals.*
—*To become familiar with the more common airborne bacterial, viral, and fungal diseases of animals, as well as other airborne pollutants that affect animals.*
—*To learn the meaning of such terms as chronic, acute, synergistic and fluorosis.*
—*To recognize how the economy is affected by air pollution effects on animals.*

Relatively little research has been done regarding air pollution effects on animals as compared to the number of studies related to the effects on man and on vegetation. Acute injuries have been reported near mines and process industrial plants during major air pollution episodes in the past. Danger to animals from accumulation of certain air pollutants has been reported as the result of laboratory experiments.

AIR POLLUTION EPISODES

The following air pollution episodes exemplify the effects of multiple pollutants in the atmosphere at high concentrations accompanied by weather inversions, where acute illness and death of animals have resulted. The Meuse Valley case of 1930 was an extreme situation where cattle became sick and had to be slaughtered. At Donora in 1948 canaries and dogs were significantly affected, although cattle, sheep, horses, and swine were not. At Poza Rica, Mexico, in 1950, canaries, chickens, cattle, pigs, and dogs became ill or died. In 1952 cattle in London had to be slaughtered because of incurable illnesses resulting from air pollution.

AIR POLLUTANTS THAT AFFECT ANIMALS

Some air pollutants cause chronic poisoning when they are emitted into the atmosphere for long periods of time. A few air pollutants, such as arsenic, fluorine, lead, molybdenum, and selenium have also caused acute injury and death. Unlike air pollution effects on humans, the route of exposure in animals is mainly through ingestion of pollutant-contaminated forage or feed.

biological aerosols

A few bacterial, fungal, and viral diseases of domestic animals attributed to airborne aerosols are listed in Tables 4–1 and 4–2.

airborne pollutants other than biological aerosols

Arsenic   Arsenical air pollution poisoning has caused deleterious effects on animals grazing near smelters processing arsenical ores. Some ill effects have occurred when livestock have grazed near grain-processing plants where flue gases containing arsenic in the coke spread arsenical air pollutants as far as 6 miles from the plant. In 1902, grazing cattle, horses, and sheep were extensively poisoned by arsenical fallout near a copper smelter in Anaconda, Montana. Patho-

TABLE 4-1   Airborne Bacterial and Fungal Diseases of Animals

Disease	Host	Causative Agent	Remarks
Bovine tuberculosis	Cattle swine, sheep, dogs, cats	Mycobacterium bovis (bacteria)	Contagious to man. Control by slaughter.
Glanders	Horses, mules	Actinobacillus (bacteria)	TB-like nodules and ulcers in respiratory tract, internal organs and on the skin. High death rate.
Aspergillosis	Birds, pigeons, ducks, chickens	Aspergillosis fumigatus (fungus)	Inhaled from grain and straw contaminated with mold. Affects pulmonary tract.
Cryptococcosis	Horses	Cryptococcus neoformans (fungus)	Causes granuloma in horse's lungs. Found in pigeon manure.
Coccidio-mycosis, valley fever, desert rheumatism	Cattle, horses, swine, dogs, wild rodents	Coccidioides immitis (fungus)	Varies in severity, symptoms of common cold.
Histoplasmosis	Domestic animals	Histoplasma capsulatum (fungus)	Causes lung lesions from inhalation of spores.

logical effects are inflammation of the respiratory and gastrointestinal tract, destruction of red blood cells, and kidney damage.

Some acute poisoning symptoms are salivation, thirst, great uneasiness, odor of garlic on the animal's breath, followed by trembling, stupor, and convulsions prior to death. Horses may have ulcers of the nose, puffiness of the eyes, dilation of the pupils, difficult breathing, and partial paralysis of the hind legs.

In chronic poisoning, arsenic appears to have a depressing effect upon the central nervous system. The animal has loss of appetite, loses weight, and may eventually become paralyzed and die.

TABLE 4-2   Airborne Viral Diseases of Animals

Disease	Host	Symptoms and Effects	Morbidity and Mortality	Control
Hog cholera	Swine	Fever, stilted gait, conjunctivitis, diarrhea	As high as 90% mortality	Immunization
Equine influenza	Horse	Fever, nasal discharge, abortion in mares	Low mortality	Immunization
Swine influenza	Swine	Exudative bronchitis	Morbidity almost 100%; mortality 2% or less	None
Feline distemper	Cat, mink, raccoon	Vomiting, diarrhea, nasal and eye discharge	Recovery usual	Immunization
Canine distemper	Dog, fox, mink	Fever, diarrhea, rhinitis	Recovery usual	Immunization
Newcastle disease	Chicken, turkey, ducks, other fowl	Coughing, sneezing, paralysis of legs, loss of egg production	Morbidity 100%; mortality 5-50%	Immunization
Infectious bronchitis	Chicken	Rales, wheezing, loss of egg production	Mortality up to 60% in chicks, negligible in older birds	Immunization

Extracted from APTD 69-30.

Cadmium Although little study has been made regarding cadmium in the environmental air, studies have been made in which cadmium capsules administered to cows for control of intestinal worms caused a decline in milk production. Pigs have been killed by administering worm medicine with small amounts of cadmium. Since cadmium is produced as a by-product when refining metals such as zinc, lead, and copper, and is used in manufacturing pesticides and fertilizers, animals grazing in close proximity to such plants may be endangered by cadmium air pollution.

Chlorine Accidental spillage is the only incidence where chlorine has been reported in the atmosphere in concentrations high enough to affect animals. However, industrial liquefaction processing, other industrial uses of chlorine, and accidental leakage during storage or transportation should be considered potential hazards.

Fluorine Excessive outputs of fluorine in phosphate fertilizer and aluminum production have been responsible for animal deaths. Cattle and sheep are the most susceptible to fluorine poisoning (fluorosis). When fed plants loaded with hydrogen fluoride, a corrosive action ensues on all body tissues.

Acute poisoning symptoms are loss of appetite and body weight, lameness, muscular weakness, and finally death. In cases of chronic poisoning, teeth become soft and mottled, a long overgrowth appears on leg bones, which causes lameness, and animals may become rubber legged and unable to stand.

Lead In Germany in 1955, cattle and horses grazing within a radius of 5 kilometers of two lead and zinc foundries became lame and had to be slaughtered. Dust samples collected in the vicinity of these sources showed lead ranging from 17 to 45 percent and zinc from 5 to 23 percent of the total weight of the samples. Brass foundries, combustion of coal and fuel oil, and auto exhausts also contribute to the lead in the atmosphere. About 97 percent of all airborne lead is put there by automobiles burning leaded gasoline.

Acute lead poisoning of animals is characterized by nervous, exciteable jerking of muscles, frothing at the mouth, delirium, stupor, and collapse. Gastric disturbance is followed by paralysis of the digestive tract and diarrhea. Chronic lead poisoning may cause damage to the blood-forming system of the body, and lead accumulates in the skeleton, kidneys, liver, pancreas, and lungs.

Mercury   Elemental mercury and most of its derivatives are protoplasmic poisons that can be lethal to animals. Russian experiments with animals indicate continuous exposure to mercury vapor above 0.3 µg/m³ of air may present a health hazard. Some organic mercury compounds are even more toxic than elemental mercury. Farm animals have been poisoned as a result of eating plants treated with mercury-containing pesticides.

The mining and refining of mercury and the use of mercury in various industries appear to be significant sources of air pollution. One incident of mercury poisoning was reported in cows and other domestic animals after fire in a nearby mercury mine.

Animals exhibit toxicity symptoms similar to man (damage to brain and nervous system accompanied by exaggerated emotional response and muscular tremors), but are more susceptible to lower concentrations of mercury.

Molybdenum   In Sweden in 1954, cattle grazing one tenth of a kilometer from a steel plant were poisoned. One died and many were affected. Pasture vegetation showed a molybdenum concentration of 230 mg/kg of dry matter, with normal levels about 1.5 mg/kg.

Ozone   Total oxidant exposures of less than 1.0 ppm from heated exhausts have been correlated with lung tissue changes in guinea pigs. Other laboratory animals have developed bronchitis, emphysema, pneumonia, and loss of fertility at levels of 1.0 ppm ozone.

Radioactive materials   Atmospheric radiation arises from material sources—rocks, soil, and cosmic rays—and from artificial sources such as nuclear explosions and nuclear industry in general. Although at present there have been no significant exposures reported from nuclear industry, projected nuclear expansion may create more problems in the future.

The effects of radiation on animals is primarily due to damage to the blood–forming centers in the bone marrow and lymph glands. Leukemia, cancer, and genetic mutation effects may result from long-term exposure to radiation levels too low to cause death.

Animals consume plants that contain radionuclides such as strontium 90, iodine 131, and cesium 137 and tend to concentrate them in their flesh and milk. Man then is subjected to radioactive contamination when he consumes the flesh and milk of these animals.

Selenium   Sources of atmospheric selenium include combustion of industrial and residential fuels, refinery waste gases and fumes, and incineration of wastes including paper products containing selenium.

Herbivorous animals have been reported killed from eating plants or foods containing toxic amounts of selenium. Ingestion of plants containing organic selenium compounds has caused chronic poisoning, known as *blind staggers.* Another type of chronic poisoning, known as *alkali disease,* has resulted from eating plants or grains containing protein-bound selenium.

Zinc   Cattle and horses have been poisoned by inhaling air contaminated by zinc and lead within 5 miles of smelters. The main effects include emaciation and swelling of limb joints, causing severe lameness that necessitates slaughter.

Table 4–3 summarizes the effects on animals of airborne pollutants other than biological aerosols.

Table 4–4 summarizes NO_2 effects on animals.

ECONOMIC SIGNIFICANCE

Air pollution affects the economy not only by killing animals; it has indirect effects such as a substantial loss due to decreased reproductivity, decreased growth, and decreased output of milk, eggs, and wool.

REFERENCES

Magill, P. L., F. R. Holden, and Charles Ackley, eds., *Air Pollution Handbook.* New York: McGraw-Hill Book Company, 1956.

U.S. Department of Health, Education, and Welfare, *Air Quality Criteria Pamphlet* (Photochemical Oxidants), PHS (NAPCA), Pub. No. AP-63. Washington, D.C.: U.S. Government Printing Office, 1970.

U.S Department of Health, Education, and Welfare, *Preliminary Air Pollution Survey—A Literature Review,* PHS (NAPCA), Pub. No. APTD 69–26. Washington, D.C.: U.S. Government Printing Office, 1969.

U.S. Environmental Protection Agency, *Environmental Lead and Public Health,* APCO, Pub. No. AP-90. Washington, D.C.: U.S. Government Printing Office, 1971.

TABLE 4-3   Other Airborne Pollutant Effects on Animals

Pollutant	Causative Agent or Source	Effects or Disease
Arsenic trioxide	Arsenical and copper smelters; arsenical pesticides	Chronic poisoning of cattle, horses, sheep
Asbestos	Laboratory experiments	Lung cancer in lab animals
Cadmium	Cadmium capsules as wormers	Cows—decline in milk Pigs—killed
Chlorine	Accidental spillage	Killed affected animals
DDT	Pesticide	Interferes with calcium deposits causing premature cracking of eggs of penguins, bald eagles.
Fluoride, hydrogen fluoride	Phosphate fertilizer and aluminum production plants	Fluorosis, corrosive action on teeth and bones. Causes rubber-legged condition in domestic stock.
Lead	Lead foundries	Lameness of cattle and horses
Mercury	Pesticides	Animals poisoned by eating forage treated with mercury-containing pesticide
Molybdenum	Steel plant	Death of cattle
Nitrogen dioxide (see Table 4-4 for NO_2 effects on lab animals)	Lab experiments	Lab animals more susceptible to pneumonia when subjected to NO_2 dosage
Ozone	Lab experiments	Bronchitis, emphysema, pneumonia. Loss of fertility of lab animals.

TABLE 4-3   (Cont'd)

Pollutant	Causative Agent or Source	Effects or Disease
Radioactive materials	Strontium-90 Iodine-131 Sesium-137	Concentrations in animal flesh and milk; food chain effects are cumulative
Selenium	Highly selenious weeds. Organic selenium compounds. Plants or grains containing protein-bound selenium.	Acute poisoning. Death of herbivorous animals. Chronic poisoning known as blind staggers. Alkali disease.
Zinc	Inhaling zinc in air near zinc smelters	Emaciation and swelling of limb joints of cattle and horses requiring slaughter

U.S. Environmental Protection Agency, *Air Quality Criteria Pamphlet* (Nitrogen Oxides), PHS (NAPCA), Pub. No. AP-84. Washington, D.C.: U.S. Government Printing Office, 1971.

RECOMMENDED FILMS

Crisis of the Environment
    TS-101-D   Preserve and Protect
    TS-101-C   Vanishing Species
    (Film strips with record, test, and manual)
    M-1707-X   Beware of the Wind (22 min)
    Available: Distribution Branch
               National Audio-Visual Center (GSA)
               Washington, D.C. 20409

QUESTIONS

1/ What is meant by chronic poisoning?
2/ How do animals usually attain chronic air pollution poisoning?

TABLE 4-4 Summary of Representative NO$_2$ Effects

| Effect | NO$_2$ Concentration | | Duration | Comment |
	ppm	$\mu g/m^3$		
Rabbits—structural changes in lung collagen	0.25	470	4 hr/day for 6 days	Still apparent 7 days after final exposure
Rats—morphological changes in lung mast cells characterized by degranulation	0.5	940	4 hr	Possibly precedes onset of acute inflammatory reaction
	1.0	1880	1 hr	
Mice—pneumonitis; alveolar distension	0.5	940	6 to 24 hr/day for 3 to 12 mo	Possibly emphysematous condition
Mice—increased susceptibility to respiratory infection	0.5	940	6 to 24 hr/day up to 12 mo	Based on mortality following challenge with K. pneumoniae

Rats—tachypnea, terminal bronchiolar hypertrophy	0.8	1504	Lifetime, continuously	Possibly pre-emphysematous lesion
Rats—bronchiolar epithelial changes, loss of cilia, reduced cytoplasmic blebbing, crystalloid inclusion bodies	0.8 to 2.0	1504 to 3760	Lifetime, continuously	
Rabbits—structural changes in lung collagen	1.0	1880	1 hr	Denaturation of structural protein suggested
Rats, monkeys—polycythemia	2.0	3760	3 wk continuously	Approximate doubling of red cell number with lesser increases in hematocrit and haemoglobin
Monkeys—tissue changes in lungs, heart, liver and kidneys	15 to 50	28,200 to 94,000	2 hr	Degree of damage directly related to concentration of NO_2

Extracted from Air Quality Criteria for Nitrogen Oxides, AP-84.

67

3/ What type animals are most affected by arsenic air pollution? What two ways may this poisoning come about?

4/ What are two bacterial airborne diseases that affect animals?

5/ What are two fungal airborne diseases that affect animals?

6/ What are two viral diseases of animals that may result from airborne pollution?

7/ What is meant by the word synergistic?

8/ What is fluorosis and how does it affect animals?

9/ What is the importance of radioactive material laboratory studies with animals?

10/ What is the difference between acute and chronic poisoning?

11/ What are some indirect economic losses resulting from effects of air pollution on animals?

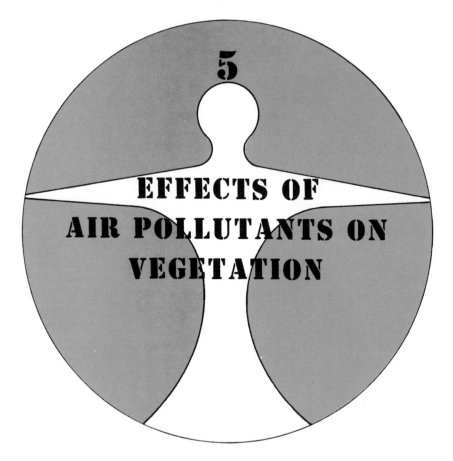

5

EFFECTS OF AIR POLLUTANTS ON VEGETATION

Student Objectives

—*To obtain an understanding of the various air pollutants that affect plants and how these may be detected.*

—*To understand the synergistic effects of multiple pollutants and other factors that complicate true evaluation of air pollution damage.*

—*To understand the basic structure and function of the plant leaf, where most air pollution effects are centered.*

—*To become familiar with such terms as necrosis, chlorosis, epinasty.*

—*To recognize the economic significance of air pollution effects on plants along with methods of control.*

A study of air pollution effects on vegetation is important not only because of economic loss from damage to the vegetation, but also because vegetation injury provides an indication of how air pollutants may eventually affect man. Observation of injury on sensitive plant species has provided a means of monitoring pollutant emissions from a source, particularly emissions of fluoride, sulfur dioxide, ozone, and peroxyacetyl nitrate (PAN). Extensive research in growing plants under controlled conditions and exposing them to various concentrations and combinations of certain chemicals permits us to interpret characteristic injuries to certain plants not only to determine the presence but even the relative concentration of aerial pollutants that cause plant injury.

A number of the most serious pollution offenders have been identified and their effects have been determined. However, investigators must use caution in diagnosing pollution injuries, since synergistic effects of multiple pollutants have been revealed. Factors such as physiological disturbances, various plant pathogens, nutrient deficiencies and plant damage due to insecticides complicate true evaluation of air pollution damage. Some environmental stresses that cause damage are marginal leaf burn due to lack of moisture, leaf spot and leaf roll from virus infection, and necrotic leaf margins damaged by low temperatures.

STRUCTURE AND ACTIVITIES OF PLANTS

A typical plant cell (Fig. 5–1) has three components: the cell wall, the protoplasm, and the nonliving inclusions within the cell. Because the cell wall is extremely thin during the formative stage, thickening with age, new growth is very susceptible to air pollution damage. The protoplasm is composed of several chemical compounds (fats, proteins, and carbohydrates), water, and the central nucleus, which contains the hereditary and reproductive mechanism. The leaf also contains the chloroplasts, which are the key structure in the photosynthesis process of food manufacture in the green plant. These plant inclusions (Fig. 5–1) are the storehouses for food and waste material.

A cross section of a leaf (Fig. 5–2) shows four principal layers: the upper epidermal cells, the palisade parenchyma, the spongy parenchyma, and the lower epidermal cells. The epidermal cells that protect the inner tissues have oval openings called stomata, bounded by guard cells. Through the stomata, gases and airborne pollutants enter the leaf. The guard cells act to close or open the stomata by changes in water content.

The mesophyll cells (palisade and spongy parenchyma) carry on

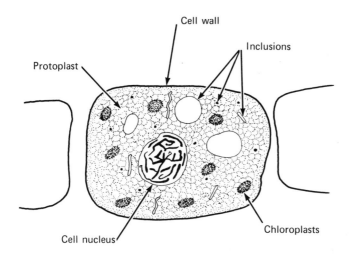

All or part of a cell may be injured by air pollutants.

Fig. 5–1 Plant cell (AP 71).

the basic food-manufacturing process. During the daytime while this photosynthesis is taking place, plants take in carbon dioxide and water, which combine to form sugar, then starch. The excess oxygen generated in this process escapes from the plant into the atmosphere. At night this exchange is reversed. At night, with lack of sunlight, photosynthesis is inactive, and respiration (energy-producing mechanism) uses oxygen as fuel to burn carbohydrates and produce energy. The plant cells then take in oxygen and release carbon dioxide to the atmosphere. The spongy layer contains veins or vascular bundles through which water, soil minerals, and food are transported to other parts of the plant. The exchange of moisture between the leaf and the atmosphere has a cooling effect on the leaf. Air pollution may effect a change in the normal water relationship, causing *plasmolysis* or breakdown of cell structure.

TYPES OF INJURY

Acute injury by air pollution to plants is a severe, visible damage to leaf tissues often associated with plasmolysis and tissue collapse. The destruction of leaf tissues or severe drying or burning is referred to as *necrosis*.

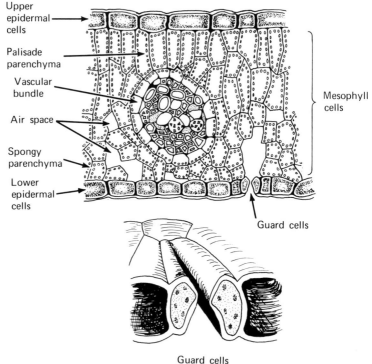

Upper epidermal cells

Palisade parenchyma

Vascular bundle

Air space

Spongy parenchyma

Lower epidermal cells

Mesophyll cells

Guard cells

Guard cells (enlarged)

Cross section of intact leaf shows air spaces within leaf that serve as passages for pollutants that may subsequently injure the leaf.

Fig. 5–2 Cross section of leaf (AP 71).

Bifacial necrosis (gradual decay) is the result of all tissues being killed on both upper and lower surfaces of the leaf.

Chronic injury, as differing from acute injury, is injury resulting from long-term exposure to low levels of pollutants and often shows up as a color change or *chlorosis* because of destruction of chlorophyl with no apparent cell damage.

Pigmented lesions may also result with dark brown, black, purple, or red spots appearing.

Growth alterations are a hidden injury not always readily detected, but may show up as growth reduction, stimulated lateral growth, decrease in apical dominance (influence of terminal buds in suppressing growth of lateral buds), twisted, droopy, or stunted structures, leaf drop (abscission), or failure of flowers to open properly. *Epinasty* is the rapid growth of the upper side of leaves, causing the leaf blade to curl under.

POLLUTANTS OF MAJOR CONCERN

Aldehydes (organic compounds that include acrolein and formaldehyde) are products of incomplete combustion of hydrocarbons and other organic materials. They are emitted into the atmosphere particularly in exhaust from automobiles. Furthermore, these aldehydes may be formed as secondary pollutants, due to photochemical reactions after entering the atmosphere. Petunias in greenhouses have suffered leaf damage when aldehyde content of the air exceeded 240 $\mu g/m^3$ for a 2-h period. Damage appeared as necrotic banding of the upper leaf surface and glazing of the lower leaf surface. Alfalfa exposed to 250 $\mu g/m^3$ of acrolein for 9–h suffered similar leaf damage.

Ammonia used as a fertilizer has caused serious damage to vegetation only as a result of accidental spills that allowed high concentrations of the gas to be carried by the winds over vegetation in the vicinity. Ammonia injury causes acute tissue collapse, with or without chlorosis.

Arsenic is used as a desiccate (drying agent) for cotton prior to machine picking and as a soil sterilant for weed control. When arsenical compounds are not used in strict compliance with instructions, they have caused damage to plants. In several cases copper and gold smelter arsenic emissions have caused damage to plants in the vicinity of the smelters.

Beryllium is commonly found as an atmospheric pollutant within the confines and in the proximity of industrial plants producing beryllium. Some investigations have determined that beryllium is toxic in levels in excess of 1 ppm in soil solution. At levels between 0.5 and 5 ppm bush-bean growth rates were reduced from normal growth rates.

Biological aerosols, disease agents borne by air, can be very harmful to plants. Some bacterial, fungal, and viral diseases are disseminated by insects, birds, animals, and water, but many are subject to airborne dispersal, especially the fungi. Table 5–1 lists airborne plant diseases.

Boron is one of the trace elements needed by plants in minute amounts, without which the growth of a plant is abnormal and death eventually results. Borates are found in some commercial fertilizers. However, boron used in large quantities as a herbicide is extremely toxic to most plants and must be used in accordance with instructions to avoid plant damage.

Carbon monoxide comes primarily from internal combustion engines. A significant impact on vegetation and associated microorganisms seems improbable. However, high levels of CO can disrupt the nitrogen fixation of free-living bacteria, as well as those nitrogen fixers found in the roots of legume plants. This results in an indirect effect on vegetation in that the plant suffers from lack of nitrogen if these bacteria are unable to convert the free nitrogen into a form usable by plants.

Chlorine is found in the atmosphere primarily in those areas where it is being used as a disinfectant, such as a swimming pool or water purification plant, or where it is involved in a chemical process and leaks from storage tanks or hydrochloric acid mists are emitted. Marginal and tip necrosis and bleaching of the foliage have occurred at concentrations of 300 to 4,500 $\mu g/m^3$.

Chromium used in metallurgical and chemical industries, as well as its presence in cement and asbestos, is believed to be the most likely source of air pollution by this element. Chromium stimulates plant growth, but can also be toxic to plants, depending on the type of plants and the amount of chromium in the soil. Chromium is used in fungicides and wood preservatives, for potato and tomato blight control, for seed sterilization, and in lawn fungicide control. Failure to

TABLE 5-1   Airborne Plant Diseases

Almond brown rot	Chestnut blight	Onion mildew
Azalea flower spot	Crown rust of oats	Potato late blight
Beet downy mildew	Downy mildew	Powdery mildew on barley
Blossom infection	Leaf spot on tulips	Stem rust of wheat and rye
Cedar rust	Loose smut of wheat	Tobacco blue mold
Apple rust	Maise rust	White pine blister rust
Leaf roll virus		

Extracted from NAPCA Pub. No. APTD 69-30.

apply proper safeguards and overdosage of chromium-containing fungicides could cause plant damage.

Ethylene is found principally in areas with high automobile density, as well as natural gas and fuel-oil heating systems and coal-burning and industrial processes. Ethlyene is a significant phytotoxicant (plant poison) and a major contributor to air pollution. It is one of the few hydrocarbons possessing the ability to injure plants without undergoing photochemical reaction with nitrogen oxides. In large metropolitan areas, concentrations in the atmosphere average 40 to 120 $\mu g/m^3$. Ethylene has a unique effect in that it does not directly attack plant tissue, but interferes with the normal action of the plant hormones and growth regulators, resulting in morphogenetic and physiological changes in the tissues. Ethylene causes general reduction in growth, stimulates lateral growth, and decreases apical dominance. Leaves may develop epinasty or show chlorosis, necrosis, or abscission.

Fluoride is emitted principally from fertilizer manufacturing, aluminum reduction, ore smelting, and ceramic manufacturing. Fluoride in concentrations as low as 0.1 part per billion (ppb) is toxic to some plants. Gaseous compounds such as hydrogen fluoride and silicone tetrafluoride are the compounds most responsible for fluoride injury to vegetation. Fluorides in either gaseous or particulate form can accumulate outside or inside plant leaves and cause leaf injury. An accumulation of a solid fluoride on a leaf surface will not injure the leaf. However, dew or light rain can produce the moisture necessary to make the fluoride soluble and readily absorbable by the tissues. Fluorides generally do not translocate from the leaf to other parts of the plant. Fluorides enter mainly through the leaf stomata and cause necrosis predominantly at the leaf tips and leaf margins, *not* normally in between the veins. Ragged leaf margins appear as necrotic tissues and are shattered and break off. With lack of moisture or nutrients such as calcium, magnesium, potassium, or phosphorus, the fluoride damage is greater. Large-seed fruits, grapes, pines, and gladioli are most sensitive to fluoride, showing up as overripening with separation of the fleshy parts of the fruit. Indirect damage by hydrogen fluoride includes fluorosis (fluoride poisoning), previously mentioned as affecting the teeth and bones of animals that have consumed forage containing fluoride in concentrations of 100 to 300 ppm or as low as 60 ppm if continuous.

Hydrogen chloride and hydrochloric acid have reportedly caused several incidents of plant damage near factories that manufacture alkali and glass, and in the vicinity of open burning of any chlorine-containing organic compounds. Injury to broadleaf plants is indicated by a marginal leaf burn that progresses basipetally (at the base of the petal) with prolonged exposure. In grasses the tips become brown colored after exposure to low concentrations. Threshold concentration appears to be 50 to 100 ppm (75,000 to 1,500,000 $\mu g/m^3$). However, in one experiment, at levels as low as 5 ppm, tomato plant leaves developed interveinal bronzing and bleaching, followed by necrosis within 72–h after exposure. Coniferous (cone-bearing) trees seem less affected than broadleaf vegetation, due possibly to the greater leaf surface area of the broadleaf vegetation. High humidity also appears to be a contributing factor to greater damage. Thus when relative humidity is increased, the rate and severity of damage suddenly increases.

Hydrogen sulfide gas is a by-product of certain industrial processes such as kraft paper mills, industrial waste-disposal ponds, sewage plants, refineries, and coke-oven plants. It has not appeared in the literature as a phytotoxic air pollutant under field conditions. However, elemental sulfur (wetable sulfur powders), calcium or sodium polysulfides, and polysulfide compounds used in fungicides possess pronounced phytotoxic properties, especially when the air temperature is high. Laboratory exposures of plants to hydrogen sulfide gas produced scorching of young shoots and leaves, with basal and marginal scorching of the next older leaves. This is the only pollutant known to affect the growing tip of sensitive plants, although peroxyacetyl nitrate (PAN) damage is generally associated with young plant tissue.

Mercury is put into the atmosphere by plants processing mercury-containing ore; it is also vented into the atmosphere by burning coal and oil and by incinerators burning paper treated with mercury during its manufacture. Mercury is mobile in the environment and once released may cycle between air, land, and water for long periods of time. Mercury damage to plants includes chlorosis, abscission of older leaves, growth reduction, and generally poor growth and development. Injury is usually restricted to greenhouse crops where mercury vapors from bichloride of mercury mixed with soil as a fungicide and mercury in antimildew paints are produced in a more confined atmosphere. Floral parts are more sensitive to mercury vapors than leaves.

Nitrogen dioxide resulting mostly from fuel combustion injures vegetation in addition to its role in producing ozone and PAN in the presence of light and hydrocarbons. Symptoms of NO_2 (nitrogen dioxide) injury appear as irregular white or brown collapsed lesions on tissue between the veins and near the leaf margin. Plants vary widely in their susceptibility to nitrogen dioxide damage. Azaleas, tobacco seedlings, lettuce, tomato seedlings, and pinto beans are a few plants that are highly sensitive at levels as low as 2.5 ppm. Low light intensity (at night) or very cloudy days seem to increase plant susceptibility to damage by nitrogen dioxide, probably due to the fact that during such periods nitrogen dioxide has a greater effect on suppression of carbon dioxide absorption, since oxides of nitrogen are not reacting with hydrocarbons to form photochemical oxidants when sunlight is not present. Suppressed growth is the most common damage. Upper leaf surface damage with water-soaked lesions appearing between veins has also been reported.

In terms of direct effects on vegetation growing in the field, the major sources of high levels of nitrogen dioxide observed resulted from accidental releases or spillage causing relatively short periods of exposure. *Ozone* is the principal oxidant produced from the photochemical reaction of hydrocarbons and nitrogen dioxide and therefore probably causes more injury to vegetation than any other air pollutant in the United States. The photolysis (chemical decomposition by radiant energy) of nitrogen dioxide and the concomitant (accompanying) removal of NO produced by reaction with HC free radicals results in the buildup of ozone. Ozone weather flecks (light spots) tobacco leaves, stipples (dots and specks) and bleaches the upper surface of pinto bean leaves, and flecks and stipples the upper surface of grape leaves. Pigment lesions are primarily on upper leaf surfaces because of injury to the palisade cells under the upper epidermal layer. Pines show chlorotic mottling on needle tips with premature needle drop. Corn leaves tend to collapse. Plants hardest hit are tomato, tobacco, bean, spinach, potato, and pine. Pinto beans and tobacco are damaged when O_3 exceeds 0.02 ppm over an 8-h period.

Peroxyacetyl nitrate (PAN), resulting from photochemical reactions of hydrocarbons and nitrous oxides, is extremely toxic to many plants, especially small plants and young leaves. PAN appears more damaging during periods of high light intensity, since sunlight provides the energy to produce PAN more readily than during low light intensity. Four hours of exposure to 15 to 20 ppb causes damage to lettuce, tomatoes, and petunias. Visible symptoms include bronzing, silvering, and glazing on lower leaf surfaces, with few symptoms or none visible on the upper surface. Banding across leaf blades in grasses is probably due to continued growth at night without PAN damage followed by daylight damage. Symptoms recorded from this type damage were yellow-green mottling and stippling of cotton leaves. Tomato, poinsettia, and pepper plants showed pronounced epinasty and twisting.

Particulates are not generally considered to be harmful to vegetation unless they are highly caustic or unless heavy deposits occur. High particulate emissions close to fruit- and vegetable-producing areas have coated these products and caused a reduction in quality or an increase in labor costs for cleaning. Lime deposits have caused reduction in plant photosynthesis. Cement dust appears to plug leaf stomata and block light needed for photosynthesis. Particulates may make plants more susceptible to pathogens through reduction in vigor and hardiness and may interfere with pollen germination.

Some *pesticides* affect plant taste and flavor. Herbicides and fungicides generally are selective, but when used carelessly or in excessive strength, they become nonselective and cause severe plant damage by killing plants or by causing defoliation, dwarfing, curling, and twisting, or a general reduction in growth. Cases have been reported where one herbicide, 2, 4-dichlorophenoxyacetic acid (2, 4-D) was used to dust rice and caused severe damage to 10,000 acres of cotton 15 to 20 miles downwind. Occasional damage has been reported in the vicinity of manufacturing plants that produce herbicides. Herbicide atrazine, insecticide parathion, and oil spray damage have been reported.

Radioactive substances are of main concern in that when they are concentrated in plants and animals, abnormal amounts are transferred to man in the food chain. Observable effects of radiation on plants range from mutations at low dose rates to growth inhibition and death at high dose rates. In general, radiation damage in plants is difficult to detect except at dose rates many times higher than those normally encountered in ambient air. Radiation cannot be detected without special instrumentation, and the biological effects are usually not evident until some time after exposure.

Selenium in small amounts is required for the growth of some plants. These plants may accumulate high concentrations of selenium and thereby become poisonous to animals that eat them. Other plants, like corn, wheat, barley, and rye, that do not require selenium can be damaged by the accumulation of small amounts of selenium.

Sulfur dioxide has been studied more intensively than any other pollutant, since it is one of the most dominant primary air pollutants in the atmosphere. SO_2 is freely emitted to the atmosphere during the combustion of fuel, especially soft coal and fuel oil of high sulfur content. SO_2 causes both acute and chronic plant injury. Acute injury is characterized by clearly marked dead tissue between the veins or on the margins of leaves. Chronic injury is marked by brownish-red, turgid or bleached white areas on the blade of the leaf. Markings on vegetation usually are found close to the source of emission.

There are many symptoms that develop on leaves of plants which resemble those due to sulfur dioxide; therefore, positive identification of SO_2 can be made only after all leaf symptoms and related evidence have been considered. This evidence includes presence of suspected sources of SO_2, the species of plants that develop markings, type of markings observed, pattern shown by the severity of the markings, and locations of occurrence. Plants are particularly sensitive to sulfur dioxide during periods of intense light, high relative humidity, adequate plant moisture, and moderate temperature. Younger, fully expanded leaves are usually most sensitive. Chronic plant injury and excessive leaf drop occur when SO_2 concentration is 85 $\mu g/m^3$ as an annual mean. An 8-h period of 860 $\mu g/m^3$ will cause injury to some plants.

Some specific symptomology (study of subjective evidence of disease or physical disturbance) is chronic undersurface silvering on cotton and alfalfa, dead areas between side veins on pinnately veined leaves, and necrotic streaks starting at the tips on grasses and needle-leafed plants. Other sensitive plants include pumpkins, squash, apples, tomatoes, and greens. Smartweed and ragweed have been used as SO_2 damage indicators.

Ozone and SO_2 together produce a synergistic action that reduces the injury threshold of leaf tissue. For instance, tobacco subjected to 0.03 ppm O_3 for 2–h produces no leaf damage. Also, 0.24 ppm SO_2 for 2–h produces no damage. However, a combined toxicant of 0.027 ppm O^3 and 0.24 ppm SO_2 caused a 38 percent leaf damage. This synergistic action is not fully understood.

Some pollutants tend to affect specific parts of the leaf or specific cells of the leaf that are most susceptible to that particular pollutant. Table 5–2 lists the major pollutants that affect vegetation, indicating what part of the leaf is affected.

ECONOMIC SIGNIFICANCE

There are basically two problems in ascertaining economic loss in crops. The first is determining how much of the crop is lost; the

TABLE 5-2   Summary of Pollutants, Sources, Symptoms, Vegetation Affected, Injury Thresholds, and Chemical Analysis

Pollutants	Source	Symptom	Type of Leaf Affected	Part of Leaf Affected	Injury Threshold(a) ppm	Injury Threshold(a) μg/m³	Sustnd Exp.
Ozone (O₃)	Photochemical reaction of hydrocarbon and nitrogen oxides from fuel combustion, refuse burning, and evaporation from petroleum products and organic solvents.	Fleck, stipple, bleaching, bleached spotting, pigmentation, growth suppression, and early abscission. Tips of conifer needles become brown and necrotic.	Old, progressing to young	Palisade	0.03	70	4 hrs.
Peroxyacetyl nitrate (PAN)	Same sources as ozone.	Glazing, silvering, or bronzing on lower surface of leaves.	Young	Spongy cells	0.01	250	6 hrs.
Nitrogen dioxide (NO₂)	High-temp. combustion of coal, oil, gas, and gasoline in power plants and internal combustion engines.	Irregular, white or brown collapsed lesion on intercostal tissue and near leaf margin.	Middle-aged	Mesophyll cells	2.5 / 1.0	4700 / 1880	4 hrs. / 21-48 hrs.

Pollutant	Sources	Symptoms	Leaf age	Tissue affected	Concentration		Duration
Sulfur dioxide (SO$_2$)	Coal, fuel oil, and petroleum.	Bleached spots, bleached areas between veins, bleached margin, chlorosis, growth suppression, early abscission, and reduction in yield.	Middle-aged	Mesophyll cells	0.3	800	8 hrs.
Hydrogen fluoride (HF)	Phosphate rock processing, aluminum industry, iron smelting, brick and ceramic works, and fiber-glass manufacturing.	Tip and margin burn, chlorosis, dwarfing, leaf abscission, and lower yield.	Mature	Epidermis and mesophyll	0.1 (ppb)	0.2	5 wks.
Chlorine (Cl$_2$)	Leaks in chlorine storage tanks, hydrochloric acid mist.	Bleaching between veins, tip and margin burn, and leaf abscission.	Mature	Epidermis and mesophyll	0.10	300	2 hrs.
Ethylene (CH$_2$)	Incomplete combustion of coal, gas, and oil for heating, and automobile and truck exhaust.	Sepal withering, leaf abnormalities; flower dropping, and failure of flower to open properly.	(Flower)	All	0.05	60	6 hrs.

Extracted from NAPCA Pub. No. AP-71.

(a) Metric equivalent based on 25°C and 760 mm mercury.

second is determining whether there are damaging growth effects. The direct economic impact is due to reduction of quality of the product, as well as quantity. Another factor to consider is how livestock are affected by eating vegetation that has been contaminated. Agricultural losses in the United States from air pollution are estimated at $500 million annually.

Indirectly, from the aesthetic viewpoint, air pollution reduces recreational value for all those who enjoy seeing healthy forests and other vegetation. Furthermore, air pollution damage to vegetation may be a precursor of erosion.

METHODS OF CONTROL

Reduction of air pollution is the best method of controlling air pollution effects on vegetation; however, there are some other effective measures that have proved successful in alleviating economic loss from air pollution damage to plants. In the case of biological aerosols damage, new wheat varieties have been developed that are resistant to rust (a type of fungal disease), and plants may be sprayed with fungicides such as copper salt mixtures to help control some of these damaging biological aerosols.

The growth of plants highly sensitive to air pollution such as orchids and gladioli has been discontinued in areas with high intensity air pollution and moved to areas where less damage is apt to occur. A specific case is the growth of orchids in the Los Angeles area, where extensive damage from ethylene caused growers to move production to rural areas.

Plants have been given a protective plastic covering or have been sprayed by various protective chemicals that help to alleviate damage by air pollutants. An example is spraying tobacco shade cloths with reducing chemicals to destroy the effect of oxidizing pollutants. Heavier fertilizer application or more frequent watering has been used to offset reduced growth rate resulting from air pollution.

REFERENCES

Agricultural Committee, *Recognition of Air Pollution Injury to Vegetation—A Pictorial Atlas*, APCA, Pub. No. TR-7. Pittsburgh, Pa.: Air Pollution Control Association, 1970.

U.S. Department of Health, Education, and Welfare, *Air Pollution Injury to Vegetation*, PHS (NAPCA), Pub. No. AP-71. Washington, D.C.: U.S. Government Printing Office, 1970.

U.S. Department of Health, Education, and Welfare, *Preliminary Air Pollution Survey—A Literature Review*, PHS (NAPCA), Pub. Nos. 69–23 through 69–49. Washington, D.C.: U.S. Government Printing Office, 1969.

RECOMMENDED FILMS

TF-102   Air Pollution and Plant Life (20 min)

TS-103   Air Pollution Effects on Vegetation (80 color slides with script)

Smog: It's Killing Our Trees (77-frame color slides with script)
Available: Distribution Branch
National Audio-Visual Center (GSA)
Washington, D.C. 20409

QUESTIONS

1/ What are some factors other than air pollution that must be considered when diagnosing pollution injury to vegetation?

2/ Define necrosis, chlorosis, and epinasty.

3/ What is the basic difference between leaf damage due to SO_2 and damage due to fluoride?

4/ What are three airborne plant diseases?

5/ What two methods have been used to reduce the effects of airborne plant diseases?

6/ What hydrocarbon air pollutant possesses the ability to injure plants without undergoing photochemical reaction with nitrogen oxides?

7/ What pollutant interferes with normal action of plant hormones and growth regulators?

8/ What pollutant may cause damage to leaves by accumulation on the outside? How does this occur?

9/ What pollutant affects the growing tips of sensitive plants?

10/ What pollutant probably causes more injury to vegetation than any other in the United States?

11/ What pollutant appears more damaging during periods of high light intensity?

12/ What pollutant causes greater damage when leaves are turgid, when humidity is high, and when sugar content of leaf is low?

13/ What are some of the yardsticks for measuring economic loss in crops due to air pollution?

14/ What are some of the methods that can be used to control air pollution effects on vegetation?

6

EFFECTS OF
AIR POLLUTANTS ON
NON-LIVING MATERIALS

Student Objectives

—*To recognize air pollutants that cause damage to materials.*
—*To understand how air pollutants damage materials.*
—*To learn the methods used to measure air pollutant effects on materials.*
—*To evaluate the economic significance of air pollution effects on materials.*

Previous chapters covered air pollution effects on man, plants, and animals—effects of a biological nature on living organisms. Air pollution also has considerable economic and aesthetic effects upon nonliving materials such as stone, brick, metals, mortar, wood, paint, electric wiring, rubber, paper, leather, textile materials, and foods.

Materials may be damaged directly by air pollutants in an abrasive action, which is more physical than chemical, or the damage may

85

be due to direct chemical attack, such as tarnishing of silver and etching of a metallic surface by an acid mist. The effect may be an indirect chemical attack, such as the absorption of sulfur oxides by leather followed by added moisture, which produces sulfuric acid that destroys the leather. Electrochemical corrosion of ferrous metals occurs when a layer of water containing air pollutants provides electrical conductivity that destroys the protective film of oxygen on the metal surface. The damage may be indirect in that after depositing soiling matter on building surfaces, the costs of removal and possibly repainting are definitely air pollution economic effects.

CONTRIBUTING FACTORS

The air pollution effect on materials is, in many cases, a combination of effects due to natural airborne pollutants, such as biological aerosols, and those man-made air pollutants of a gaseous and particulate nature. The intensity of effects of these natural and man-made pollutants is also weather dependent. That is, a certain level of pollution of sulfur oxides will not be as corrosive on metals in a dry climate as it will be in a humid climate where sulfur oxides combine with water to produce sulfuric acid, a highly corrosive chemical. Wind is another contributing factor, since airborne particulates become more abrasive when driven into building surfaces at high rates of velocity. Temperature is a controlling factor in most chemical reactions, and sunlight can cause direct deterioration of certain materials.

EFFECTS ON SPECIFIC MATERIALS

Buildings are affected by the abrasive action of particulates that cause physical erosion and by oily particles and other soiling materials that become embedded in building surfaces. As an indirect result, further deterioration is caused when sandblasting of stone surfaces is required to remove soilants or when paint must be scraped from buildings prior to repairing. Indirect chemical action also causes deterioration, particularly with building stones such as limestone, sandstone, or marble, which contain calcium and magnesium carbonate. Sulfuric acid from SO_x air pollution in reaction with moisture and other materials forms a loose surface that flakes off and also increases soiling effects. Table 6–1 lists specific air pollutants, type of effects, and materials affected.

Metals display corrosion effects which demonstrate that high sulfation values go hand in hand with high corrosion values. In other words, with more sulfur oxides in the air, metals tend to corrode more rapidly. However, there are also synergistic effects between sulfur oxide levels, particulate levels, and moisture content of the air. At SO_2 levels of 345 $\mu g/m^3$ combined with high particulate levels, the corrosion rate of steel is increased 50 percent or more. Also, corrosion of steel and zinc panels occurs at an accelerated rate when particulate concentration ranges from 60 to 180 $\mu g/m^3$ (annual geometric mean) in the presence of SO_2 and moisture.

Carbon in particulate matter sets up an electrochemical cell with metal, and when water containing SO_x is absorbed, electrolytic corrosion occurs. Aluminum and copper form protective surface films. Zinc also forms a surface film of basic carbonate that protects iron surfaces. In the presence of SO_x, the zinc basic carbonate is dissolved and the protective action of zinc is lost. Corrosion of galvanized metals is also faster in SO_x polluted atmospheres due to this same electrolytic corrosion effect.

Electric contacts, usually made of copper or silver, tarnish, and the resulting corrosion film acts as insulation. This insulation reduces the free flow of electricity across the contacts and may even cause a short with resultant power failure. Gold is less susceptible to corrosion and would be used more often except for the fact that gold is more expensive than silver or copper. Solid-state devices and miniaturized circuitry add new problems, since only minute amounts of corrosion may impede free flow of electricity more readily than with larger electric contacts. Computers, which use solid-state devices and miniaturized circuitry, must be protected from air pollution corrosion by placement in rooms kept free from air pollutants.

SO_x is the major corrosive gaseous pollutant because it is more prevalent among the gaseous pollutants and the element sulfur is a highly reactive element.

Saprophytic bacteria can grow on the surface of many inanimate materials when there is high humidity. Miniaturized electronic circuits can be damaged by fungi unless coated with varnish-containing fungicides.

Paints containing lead pigments darken through the formation of lead sulfide when H_2S is present as a pollutant. A white house turns a dirty gray and other colors fade. Zinc and titanium-based pigments help reduce this problem, since these elements are less reactive with

TABLE 6-1   Air Pollution Effects on Materials

Pollutant	Type of Effect	Materials Affected
Ammonia	Deterioration associated with SO_2 and moisture	Damage to varnish and paint surfaces, discolors fabrics
Biological aerosols		
Saprophytic bacteria	Deterioration of surfaces and food spoilage	All material surfaces in contact with air, most unprotected food
Fungi	Deterioration	Electronic circuits, leather
Carbon dioxide	Deterioration caused by combination of CO_2 and moisture to form carbonic acid	Building stone
Chlorine	Corrosion and discoloration	Metals, paints, textiles
Chromium	Corrosion in form of chromic acid, discoloration	Metals, paint, building materials, paper, textiles
Hydrochloric acid	Corrosion	Most metals and alloys
Hydrogen fluoride	Etching	Glass and metals
Hydrogen sulfide	Discoloration and tarnishing	Paint (especially lead based), copper, zinc, silve
Iron	Stain in form of iron oxide, soiling	Paint and other materials textiles
Manganese	Soilant, especially near ferromanganese plant	Most materials, textiles
Nitrogen oxides	Cause dyes to fade and whites to turn yellow	Textiles
Odor-producing pollutants	Odors that cling to skin, hair, clothing (especially bad near soap plant)	Requires more frequent laundering and dry cleaning of clothing

TABLE 6-1 (Cont'd)

Pollutant	Type of Effect	Materials Affected
Ozone-organic oxidants	Deterioration Fading of dyes	Cracking of rubber Textiles
Particulates	Abrasion and corrosion when combined with gaseous pollutants	Most metals, paint, textiles
Phosphorous	Corrosion in form of phosphoric acid	Most materials
Sulfur oxides	Corrosion	Steel, zinc, electrical equipment, limestone, roofing slate, mortar, statues, textiles, leather, book bindings
	Electrochemical deterioration	Iron, aluminum, copper, silver, building materials, leather, paper, textiles

sulfur than lead. The major cause of soiled paint surfaces is the deposition of particulate material combined with the abrasive action of wind and other adverse weather. Organic constituents of protective paint coatings are also subject to microbial attack and damage. Paint formulas with zinc, titanium, and tin reduce mildew and fungi on painted surfaces.

Rubber cracking has been traced to the chemical effects of ozone, which is especially deleterious in areas such as Los Angeles, where high-density auto exhaust emissions combine with the high-intensity sunlight and with frequent weather inversions. Antioxidants (substances that inhibit oxidation) have been added to rubber formulations to provide some protection.

Paper exposed to sulfur dioxide in the atmosphere turns yellow and brittle because paper tends to react like a sponge in the absorption of water, which combines with sulfur oxides to form sulfuric acid.

Chromium in the form of chromic acid also turns paper brittle and yellow.

Leather upholstery and leather bookbindings become brittle from absorption of sulfur dioxide. Under very hot and humid conditions, fungi and some saprophytic bacteria also cause deterioration of leather goods because of their affinity for a moist media in which to live.

Textiles have reduced wear life due to abrasive action from particulate pollutants, chemical reaction from gaseous pollutants, and abrasive action due to the extra washing and dry cleaning required to remove soilants. Man-made fibers show increased soiling tendencies over cotton because of their hydrophobic (lack of affinity for water) nature and static charges. SO_x and other acid aerosols in the atmosphere cause increased runs in nylon stockings and loss of strength in cotton curtains and drapes. No_x, SO_2, and ozone cause dyes to fade and white fabrics to turn yellow. Ammonia, chlorine, chromium, iron, and manganese air pollutants have been implicated in textile damage. Each of these effects is specific in nature due to the chemical and electrochemical effects of the specific material under consideration and its reaction to the given pollutant.

METHODS OF MEASUREMENT

The *effects package* shown in Fig. 6–1 is a device designed to display various types of materials for subjection to atmospheric pollution. The amount of damage caused by air pollution to these materials can then be evaluated.

To correlate the amount of damage with other contributing factors, other atmospheric air pollution sampling devices as well as meteorological sensing devices should be located in the vicinity of the effects package. Particulates and sulfur oxides are two of the most important air pollutants that have synergistic effects in the production of effects on materials. Therefore, the effects package has three attachments: a *dustfall bucket* and *sticky paper* for collection of particulates and a *lead peroxide candle* for collection of sulfur oxides. The large light-colored object on top of the effects package is the dustfall bucket. The louvered dark object on top is the lead candle with the sticky tape inside the candle.

In Chapter 7 we discuss these three sampling devices along with other particulate and gaseous sampling devices. Meteorological sensing devices are covered in Chapter 12.

Fig. 6–1 Effects package (Silver-Top Manufacturing Co., Inc.).

The components of the effects package designed to measure effects on materials include nylon panels and steel and zinc plates attached to the outside of the package and a silver plate, rubber panels, and cotton panels enclosed inside the package shelter. The shelter has louvered doors that allow atmospheric air to penetrate inside the shelter to register air pollution effects on the items placed inside.

Photochemical effects on monofilament *nylon* are tested by stretching nylon on a standard polaroid slide mount for 30 days and attaching this panel to the raised bracket on top of the effects package. The panel is then examined under a microscope to determine the number of broken fibers resulting from the 30-day exposure.

The *steel* and *zinc* plates are fastened to the metal bar on the lower edge of the top of the shelter. These plates are analyzed to determine corrosion caused by air pollution by weighing the plates initially before exposure, then weighing them again after exposure of 1 year to determine loss of weight. In addition to this gravimetric or weight analysis, they are inspected for pitting and other disfiguration. Gravimetric results are expressed as milligrams per square centimeter per day (mg/cm²/day).

The *silver* plate mounted inside the shelter is analyzed by measuring tarnish caused by air pollution as a measure of reflectance and is expressed as percentage of *decrease* of *luminance apparent reflectance* (LAR).

Unvulcanized *rubber strips* are hung inside the shelter to determine rubber cracking effects of ozone in the atmosphere. A 360-g weight is attached to the strips to establish stress on the rubber. After 7 days of exposure, the strip is cut lengthwise down the center. The cross section of nine consecutive cracks near the center of the strip is measured with the aid of a microscope, and the average depth in millimeters is computed. This provides an easy way to keep track of total oxides in the air.

Dyed cotton fabric panels are used to determine fading of dyes due to chemicals in the atmosphere. These panels are exposed inside the shelter to prevent exposure to sunlight because of the obvious effects it would have on fading. A photoelectric color instrument that measures incandescent light reflected at 45° from cloth samples compares the fabrics before and after a 90-day exposure to obtain total color difference or fading reading.

ECONOMIC SIGNIFICANCE

Economic costs of material damage include lowered performance plus the necessity of using more resistant and more expensive raw materials. Material damage effects are recognized by quality-conscious manufacturers who offer steel, dyes, and rubber products that resist the effects of atmospheric pollutants. Most of the deterioration of materials by air pollution goes unnoticed because it cannot be distinguished from what might be called normal or natural deterioration. However, material damage has been evaluated at $65/person/year. This includes cleaning costs, repair and replacement costs, overdesign, reduced property values, and unaesthetic appearance of damaged materials.

REFERENCES

U.S. Department of Health, Education, and Welfare, *Air Quality Criteria Pamphlet* (Particulates), PHS (NAPCA), Pub. No. AP-49, and *Air Quality Criteria Pamphlet* (Sulfur Oxides), PHS (NAPCA), Pub. No. AP-50. Washington, D.C.: U.S. Government Printing Office, 1969.

U.S Department of Health, Education, and Welfare, *Preliminary Air Pollution Survey—A Literature Review*, PHS (NAPCA), Pub Nos. APTD 69–23 through 69–49. Washington, D.C.: U.S. Government Printing Office, 1969.

U.S. Department of Health, Education, and Welfare, *Selected Methods for the Measurement of Air Pollutants*, PHS (NAPCA), Pub. No. 999–AP-11. Washington, D.C.: U.S. Government Printing Office, 1965.

QUESTIONS

1/ What are some of the types of damage to materials by air pollution?

2/ What are some of the factors that influence deterioration of materials?

3/ What are two damaging effects on materials caused by biological aerosols?

4/ What two pollutants have the most effect on building-material damage?

5/ What effect does carbon have on corrosion of metals?

6/ What electrical-contact metals are most susceptible to corrosion?

7/ What pollutant exerts the worst effect on paint?

8/ What is the major pollutant that causes rubber to crack?

9/ What two pollutants cause paper to deteriorate?

10/ In what way do textiles become damaged by air pollutants?

11/ In what ways are damage to materials measured? How are results expressed?

12/ What is meant in reference to "overdesign" related to economic significance of materials damage?

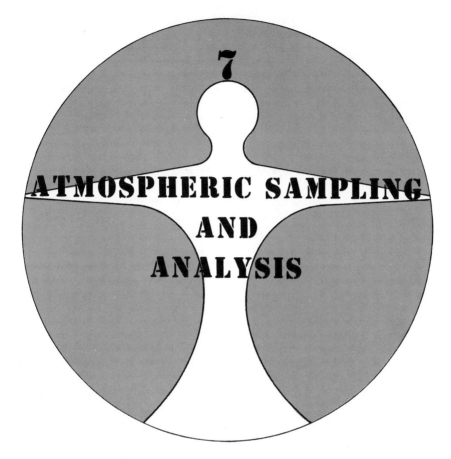

7

ATMOSPHERIC SAMPLING AND ANALYSIS

Student Objectives

—*To learn the difference between atmospheric sampling and source sampling.*

—*To develop an understanding of atmospheric sampling procedures.*

—*To become aware of the various collection techniques and sampling devices used for particulate and gaseous pollutants.*

—*To learn the various analytical procedures used to determine pollutant concentrations obtained from collection and sampling devices.*

—*To become familiar with recommended units for expressing air pollution data.*

95

There are two sampling approaches for noting the presence of air pollution. *Source sampling* obtains the pollutant count of a particular source, whereas *atmospheric sampling* deals with the pollutants within the total air mass surrounding the earth. Atmospheric sampling is the subject of this chapter.

PURPOSE OF ATMOSPHERIC SAMPLING

The purpose of atmospheric sampling is to develop air quality criteria, which are the basis for setting air quality standards. Monitoring stations collect data for determining if these standards are met. Because federal, state and local control agencies operate standard air-sampling equipment and perform standard analysis, their data are readily comparable. In summary, atmospheric sampling measures pollution, provides control data, and renders background information for use in trend evaluation and source detection.

SAMPLING PROCEDURES

The complete sampling procedure is based on the pollutant sampled, the techniques used in collecting the pollutant, the device chosen (depending upon the technique), and the method of analysis (related to the device used). Atmospheric sampling is related to analysis in that a volume of air [e.g., 1 cubic meter (m^3)] is collected. This volume of air is analyzed to determine quantity of air pollutant collected, which is measured in micrograms (μg). The pollution concentration in the air sample is then expressed in micrograms per cubic meter ($\mu g/m^3$).

factors to be considered in sampling

The *location of the sampling equipment* is decided upon first, and then the length or *duration* of the sampling period is established according to the purpose of the test. The purpose can be to find the average pollution concentrations in a given area or the purpose can be to define the peak periods of pollution in the area. A 24-h sampling period is often prescribed to determine average pollutant concentrations, whereas peak periods may be ascertained by using a continuous recorder or by taking periodic samples with a sequential sampler equipped with a timing mechanism. The sequence of samples may be every hour, every two hours, or longer sequences.

Once the location of the sampling station and the sampling duration are determined, the *sample size* is figured. This must always be large enough to provide statistical accuracy. As an example, where air pollution is high, samples of 1-h duration may provide a measurable amount of pollutant for statistical accuracy. On the other hand, in an area of relatively low air pollution, it may be necessary to collect samples over a 24-h period before obtaining a measurable amount of pollutant. This in turn may require a change in duration of sampling. The *rate of sampling* is also related to sample size. This rate is based on the contact time required between the pollutant and the absorbing or adsorbing reagent used to detect pollutant concentrations. When the sample volume is increased, a pressure drop results, and the reagent increases its percentage of penetration. Rate of sample flow is varied until the best efficiency is determined.

Some other factors may also be considered. For instance, gaseous interference of NO_2 concentrations may affect sampling for SO_2, requiring counteracting reagents to be added; the Jacob–Hochheiser NO_2 procedure, although less efficient than the Saltzman procedure, may be preferred to prevent time-lapse deterioration effects on color; equipment available may dictate use of the Saltzman procedure for NO_2 sampling where more accuracy can be obtained and color deterioration is not a factor; the units used in expressing results may dictate a specific-type analyzer to obtain comparability with other data gathered; and limited availability of samples and laboratory analysis facilities may dictate specific sampling techniques.

sampling train components

The fundamental components of a sampling device include a sampling train, which embodies an *air mover*, a *flow-measuring device*, and a *sample collection mechanism* with a *contaminant detector* that provides a means for analyzing the sample. The air mover creates a flow of air that will allow the contaminants in the air to be captured in the sample collection mechanism for analysis. Some examples of air movers are vacuum pumps, ejectors or aspirators, liquid displacers, and evacuated flasks. The latter two combine the functions of air mover and collection mechanism.

All sampling trains must have a flow-measuring device to determine the amount of air passing through the train during a given period of time. This device must be *calibrated* against a very accurate meter accepted as a standard meter.

Sample collection mechanisms vary considerably, but all must be designed to collect the air sample and provide a means for detecting the amount of pollutant contained in the sample of air collected.

PARTICULATES

Pollutants are classified as particulates or gaseous. The collecting devices and the techniques used are varied and will be considered separately. Particulates will be considered first.

Particulates vary in physical properties such as size, density, shape, and specific gravity. These properties affect the time the particle remains in the atmosphere and also the sampling method chosen. If a pollutant is free floating, it is a *suspended particulate*; if it separates out of the air, it is a *settleable particulate*. For example, oil soot, being small and of low density, remains in the atmosphere for prolonged periods; thus, it is a suspended particulate and is sampled according to suspended particulate procedures. Pollens also have a low density and a shape that tends to keep them airborne. Fly ash and dust, however, are high in density and are medium to large in size and therefore settle readily. This distinguishes them as settleable particulates. Two other particulate designations are *radioactive particulates*, such as gross beta particulates, and *total particulates*, which includes all measurable particulates.

collection techniques and sampling devices

Collection techniques are determined by the particulate involved. After the pollutant is identified and the collection technique is decided upon, a sampling device can be chosen. In some instances there are several instruments available for performing each technique. After presentation of each collection technique, sampling devices applying that technique will be described. The operation of those most frequently used will be discussed and others will only be listed. Figures 7–1 through 7–13 show these particulate sampling devices, and a summary of the particulate samplers is listed in Table 7–1.

The **gravity technique** is used to collect settleable particulates that settle out of the atmosphere due to gravitational pull. Devices that use this technique are referred to as dustfall sampling instruments. This equipment is inexpensive, simple to operate, and requires only a weight determination for analysis. Dustfall sampling devices are used to find average pollution concentration in a given area, but cannot determine peak pollution periods as well as other devices. Furthermore, only a small air sample is collected and used to represent a much larger air mass. Some factors influencing analytical findings are particles added by washout (scrubbing of the atmosphere by precipitation) and rainout (formation or condensation of moisture around a particle), dust from heavy traffic, and materials redistributed by wind and air currents. Information of this type should be reported along with analytical results.

Dustfall sampling may be done with a *dustfall brush* or *cellophane tape* coated with a material that captures dust on its sticky surface, but is most often done with a *dustfall bucket* (shown in Fig. 7–1 as the light-colored object). The bucket is approximately 8½ in. high and contains water that collects the particulates. During warm weather, ammonium chloride is added to the water as an algicide to prevent growth of algae in the water, which would add to the weight of particulates being collected. During winter weather, isopropyl alcohol is added as an antifreeze to prevent formation of ice, which would prevent pollutant collection by the water.

Dust is rinsed from the bucket, evaporated to dryness, and then weighed in milligrams. The amount of air sampled is based on the measurement of the area of the bucket. The standard collection period is 30 days. Results are reported in milligrams per cubic meter per 30 days.

The *filtration technique* collects suspended particulates that do not settle out of the air. Particles are removed from an air sample by a suction apparatus (e.g., a vacuum pump) and are deposited on a porous filter. Filtration may also be used to collect radioactive particulates. The *high-volume sampler* (Figs. 7–2 and 7–3) is most frequently used when sampling for suspended particulates. This device is sheltered to prevent collection of settleable particulates and a high-volume constant-flow pump draws air under the eaves on the shelter and through a glass-fiber filter where the particulates are deposited. The unit is carefully calibrated against a standard meter to determine the quantity of airflow in cubic meters collected over a 24-h period. The sampler allows for a large volume of air (2,016 m^3) to be filtered in a short period of time. The filter is weighed prior to collection and after collection to determine weight in micrograms. Results are reported in micrograms per cubic meter.

The *paper tape sampler* is another filtration instrument used for suspended particulates. It is most suitably adapted to collect fine or

Fig. 7–1 Dustfall bucket and lead peroxide candle (Courtesy of Research Appliance Co.).

Fig. 7–2 High-volume sampler in shelter (Courtesy of Research Appliance Co.).

soiling materials. It is particularly advantageous because it automatically gives a large number of readings per day, estimating the soiling potential of certain pollutants. Also, a wide variety of sampling periods can be selected. (Figure 7–4 shows a front view of the tape sampler.)

The paper tape sampler contains a pump that vacuums particles from the air and deposits them on a cellulosic tape filter. The tape moves automatically at predetermined intervals, usually every 2-h, to where another sample is collected. The timer can be run over a 24-h period. Readings are made from the tape by a transmissometer or densitometer (Fig. 7–5). (See absorption of radiation methods of analysis later in this chapter.)

In the *inertial technique* total particulates are collected. In this technique a polluted airstream containing particulates is drawn into a sampler where obstacles are placed across the path of the airstream. The obstacle causes the airstream to change direction, but the particles continue to travel in the initial direction and collide with the obstacle. If the obstacle has an adhesive substance, the particulates are *impacted*

Fig. 7–3 High-volume sampler motor and filter with attached recorder (Courtesy of OMD-APCO-EPA).

Fig. 7–4 Paper tape sampler (Courtesy of Research Appliance Co.).

on its surface. If the obstacle is immersed in a fluid, the particulates are collected by *impingement* in the liquid. If the inertial device is designed to force the airstream rapidly through a circular path, the particulates are drawn outward from the center and separated by *centrifugal* force.

Some devices using the *inertial-impaction technique* are used to collect specific types of particulates. The *Durham sampler* collects pollen on a vaseline-coated microscope slide, which may then be examined under a microscope. The *rotorod sampler* (Fig. 7–6) and the *Hirst spore trap sampler* (Fig. 7–7) collect pollen or spores and can be oriented to detect the direction from which these particulates are

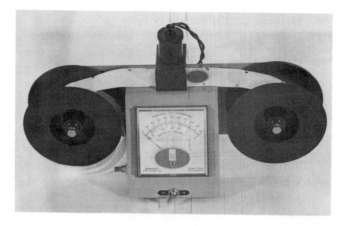

Fig. 7–5 Transmissometer (Courtesy of Research Appliance Co.).

Fig. 7–6 Rotorod (Courtesy of OMD-APCO-EPA).
Fig. 7–7 Hirst spore trap (Courtesy of OMD-APCO-EPA).

blown by the wind and can be oriented as to time of heaviest pollution. The *Andersen sampler* (Fig. 7–8), which collects bacteria or other airborne particulates, is constructed with a series of eight stainless-steel plates, designed to simulate the human respiratory system, that collect particles comparable to those which would have penetrated the respiratory tract through the pharynx down to the alveoli. The *cascade impactor* (Fig. 7–9) collects total particulates for microscopic sizing. *Sticky tape* (Fig. 7–10) wrapped around a jar relies on the wind to give directional count of total particulates, and the Gruber *particle counter* (Fig. 7–30) is used to analyze this sticky tape.

Some devices using the *inertial-impingement technique* are the *Greenburg–Smith impinger* (Fig. 7–11, left) and the *midget impingers* (Fig. 7–11, center).

The Greenburg–Smith impinger is a glass cylinder with a smaller concentric glass-tube insert. At the bottom of the tube, a glass jet and a glass impingement structure are immersed in collecting fluid. It is good for capturing dust, mist, fumes, soluble gases, particulates, and unsoluble particulates greater than 2 μ.

A sampling device using the *centrifugal-inertial technique* is the

Fig. 7–8 Andersen particulate sampler (Courtesy of Andersen 2000, Inc.).
Fig. 7–9 Cascade impactor (Courtesy of C. F. Casella and Co., Ltd.).

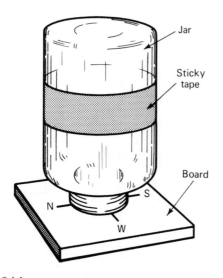

Fig. 7–10 Sticky tape sampler (Courtesy of OMD-APCO-EPA).

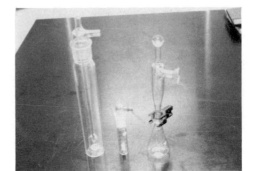

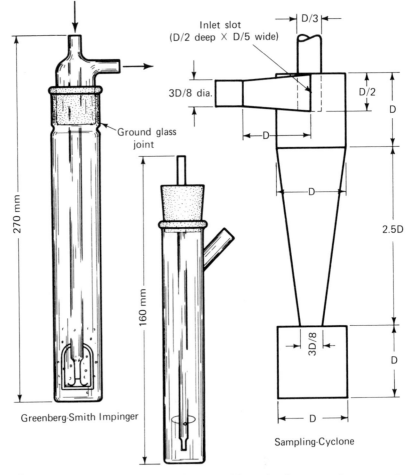

Inlet slot
(D/2 deep × D/5 wide)

D/3

3D/8 dia.

D/2

D

Ground glass
joint

270 mm

D

D

2.5D

160 mm

3D/8

D

Greenberg-Smith Impinger

D

Sampling-Cyclone

Fig. 7–11 Greenberg-Smith impinger, midget impinger, cyclone sampler (Courtesy of OMD-APCO-EPA).

cyclone sampler. It accumulates particles greater than 5 μ in diameter (Fig. 7–11, right).

The **precipitation technique** is divided into thermal and electrostatic precipitation. Thermal precipitation uses a heated wire to drive radioactive particulates by thermal convection and molecular bombardment out of gaseous streams onto a cold collecting surface. Particles 0.01 to 10 μ adhere best to a collecting surface. Electrostatic precipitation uses an electric charge to force radioactive particulates to migrate out of the airstream onto a collecting surface. This technique is good for microscopic and chemical counting and sizing of radioactive particles. It is most effective for particles 0.01 μ to 10 μ that can be impressed by an electric charge. This technique cannot be used in the presence of explosive gases, since the heated wire or the electric charge would cause an explosion.

Sampling devices using the *precipitation technique* include the *electrostatic precipitator* (Fig. 7–12), used for collection of particulates and for studying type and shape of radioactive particulates, and the *thermal precipitator* (Fig. 7–13). For a summation of particulate sampling equipment, see Table 7–1.

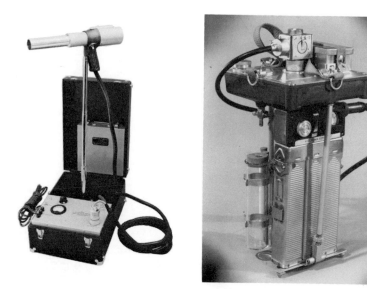

Fig. 7–12 Electrostatic precipitator (Courtesy of Mine Safety Appliance Co.)
Fig. 7–13 Thermal precipitator (Courtesy of Mine Safety Appliance Co.).

TABLE 7-1   Particulate Sampling Equipment

Sampling Device	Collection Technique	Pollutant to Be Collected	Method of Analysis
Dust fall bucket (Fig. 7-1), brush	Gravity: 30 days	Settleable particulates	Gravimetric
High volume sampler (Figs. 7-2 and 7-3)	Filtration: 24 hours	Suspended particulates (also organic, inorganic and radio-active compounds)	Gravimetric
Paper tape sampler (Fig. 7-4)	Filtration: 2 hours	Suspended particulates (fine soiling matter)	Transmissometer (Fig. 7-5), densitometer
Durham Sampler Rotorod (Fig. 7-6) Hirst Spore Trap (Fig. 7-7)	Inertial (impaction)	Pollen, spores	Count
Sticky tape (Fig. 7-10)	Inertial (impaction)	Total particulates	Gruber Particle Comparator (Fig. 7-30)

Andersen Sampler (Fig. 7-8)	Inertial (impaction)	Bacteria–particulates	Count
Cascade Sampler (Fig. 7-9)	Inertial (impaction)	Total particulates	Microscopic sizing
Greenberg-Smith (Fig. 7-11)	Inertial (impingement)	Total particulates	Gravimetric
Midget Impinger (Fig. 7-11)	Inertial (impingement)	Total particulates	Gravimetric
Cyclone Sampler (Fig. 7-11)	Inertial (centrifugal separation)	Particulates of 5 microns in diameter	Gravimetric
Electrostatic precipitator (Fig. 7-12)	Precipitation	Radioactive particulates	Type study—not quantity
Thermal precipitator (Fig. 7-13)	Precipitation	Total particulates	Gravimetric

GASEOUS POLLUTANTS

Pollutants are classified as particulates or gaseous. The previous paragraphs have covered collection techniques for collecting particulates along with examples of some sampling devices. The following paragraphs cover techniques and devices most commonly used for the collection of gaseous pollutants.

collection techniques and sampling devices

The four basic techniques for collecting gaseous pollutants are absorption, adsorption, condensation (freeze out), and grab sampling.

Absorption sampling is the process by which a gaseous contaminant in air is removed by bringing the contaminant into close contact with a liquid chemical with which it will react to form a nongaseous substance. A standard chemical solution is prescribed for each gaseous contaminant to be collected. In this technique it is important to assure that all the gaseous contaminant comes in close contact with the chemical absorbing solution and is retained long enough to allow complete reaction to occur. This is to ensure a high collection efficiency.

To assist in obtaining the highest collection efficiency, the following points are considered: the smaller the size of bubble produced, as the air sample containing the pollutant passes through the absorbing solution, the greater the contact between gas and liquid; proper size of collection device and/or rate of sample flow must be provided to allow sufficient contact time to capture the pollutant; the absorbing solution concentration must be based on expected concentrations of the contaminant being collected; therefore, to be on the safe side, an excess of the reactant in the absorbing solution is preferable to ensure collection of all the contaminant should heavier pollution be encountered; since the period of contact between gaseous contaminant and liquid absorbing reagent is very short, an absorbing reagent is desired that provides an instantaneous reaction.

One device using the absorption principle is the multiple-gas sampler, commonly referred to as the *twenty-four-hour bubbler* (Fig. 7-14). This sampler is designed to sample air concurrently for a maximum of five different gases using separate suitable collecting solutions in each of the bubblers. A vacuum pump draws air through five separate trains operating in parallel between inlet and outlet manifolds. Airflow is controlled by a calibrated limiting orifice. A thermostatically controlled heater maintains a constant temperature inside the sampler. By using appropriate-type bubblers and absorbing rea-

Fig. 7–14 Twenty-four hour bubbler (Courtesy of Research Appliance Co.).

gents for each gas, samples can be collected for SO_2, NO_2 H_2S, NH_3, and aldehydes. Results are expressed in micrograms per cubic meter.

Another device using the absorption principle is the *sequential sampler* (Fig. 7–15), which is another variation of the bubbler device. It has a timing device and 12 separate bubblers, which can be timed to collect samples in sequence for varying periods of time (e.g., collect a 2-h sample, then switch to next bubbler, accumulating sequential 2-h samples over a 24-h period). This device allows peak periods of gaseous pollution to be ascertained. Figure 7–16 shows some typical *fritted glass absorbers* used in absorption trains. This fritted glass bubbler produces the small bubbles that provide better gas absorption, whereas the impinger-type insert is used when collecting particulates.

The *lead peroxide candle* (Fig. 7–1) method for estimating sulfur dioxide concentration in the atmosphere is a simple method that applies the absorption technique. In this technique a gaseous pollutant is absorbed in a hygroscopic solid rather than in a liquid. A gauze soaked with paste containing lead peroxide is wrapped around a porcelain cylinder. The lead peroxide reacts with sulfur dioxide to form lead sulfate. The amount of sulfation occurring over a 30-day period is measured by gravimetric analysis or by measuring the discoloration effects of the gauze.

Some continuous-monitoring equipment discussed later in the chapter also uses the absorption principle.

Adsorption sampling utilizes the phenomenon by which gases are attracted to the surface of a solid and retained there. The total amount of the gaseous pollutant adsorbed is related to the surface area

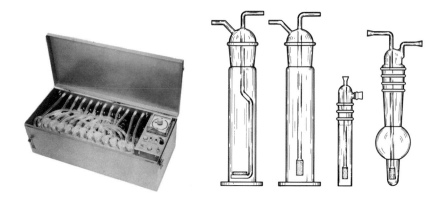

Fig. 7–15 Sequential sampler (Courtesy of Bendix-UNICO).
Fig. 7–16 Fritted glass bubblers or absorbers (Courtesy of OMD-APCO-EPA).

of the adsorbent, the pressure and the temperature maintained in the sampling train (since these two factors affect volume and thus affect concentration), and the physical and chemical characteristics of the adsorbent used. Various adsorbents are discussed next.

Activated carbon prepared from peach pits or coconut shells is one of the best materials used as an adsorbent. Various hydrocarbons on the surface of charcoal-like materials are oxidized during the activation process. This oxidation is usually accomplished by heating the char at a specific temperature for a given length of time. Some gases readily adsorbed by activated charcoal are NH_3, NO_x, CO, and CO_2.

Silica gel prepared by the coagulation of a colloidal solution of silicic acid into a hard glossy form of silicon dioxide is highly porous and is effective in adsorbing H_2S, SO_2, and H_2O.

Activated alumina is a granular adsorbent found in some tube detectors, consisting mostly of a highly porous aluminum oxide used with an indicating chemical that provides a color-forming reaction that is measurable.

A *molecular sieve*, which is a synthetic sodium or calcium aluminosilicate of very high porosity, has been used to adsorb CO_2, H_2S, SO_2, NH_3, and acetylene.

Some advantages of the adsorption principle over the absorption principle are

1/ Pollutants are rapidly detected by the color-forming reaction.

2/ The sample collected can be transported in solid form.

3/ It gives a complete collection of highly volatile gases.

4/ It provides practical temperature conditions for collection.

5/ Contaminants can be collected at high concentrations.

A *Universal Tester Kit* (Fig. 7–17) produced by Mine Safety Appliance Company uses the adsorption principle for detecting a wide variety of toxic gases, vapors, mists, and selected dusts. A simple plunger pump is used to draw an air sample through an indicator or detector tube or a specific chemically impregnated filter paper. Certain toxic compounds may be separated by application of heat, thereby releasing a chemical whose concentration can be related to the toxicant of the original compound. Heat is provided by a battery-operated pyrolizer (Fig. 7–18).

The National Bureau of Standards prescribes a CO detector tube containing silica gel impregnated with H_2SO_4 solutions of ammonium

Fig. 7–17 Universal tester (Courtesy of Mine Safety Appliance Co.).

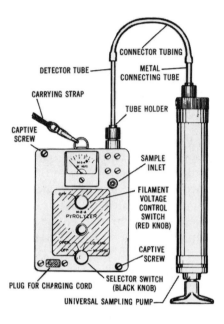

Fig. 7–18 Pyrolizer (Courtesy of Mine Safety Appliance Co.).

molybdate and palladous sulfate. As air passes through the tube, the color changes from yellow to blue-green and the amount of CO is determined by matching against a comparative color chart [adsorption CO detector (Fig. 7–19)].

Freeze-out or condensation sampling is used to collect hydrocarbons, radioactive gases, and other insoluble or nonreactive vapors. Air pollutants existing as gases can be trapped or removed by the freeze-out or condensation method. Trapping implies collecting a pollutant and removal implies freeing unwanted gas contaminants from the gas stream.

The mechanics of the freeze-out condensation process are as follows: air is drawn through collection chambers with progressively lower temperatures; if the chamber temperature is equal to or less than the boiling point of the gas, the gas will change into a liquid. This liquid or condensate is then collected in the chamber where the phase change from gas to liquid occurs. Sampling trains, using the freeze-out principle, have the ability to collect several gases at the same time when several collection chambers are provided.

A typical freeze-out sampling train (Fig. 7–20) contains five glass traps connected by ground-glass ball-and-socket joints. These traps

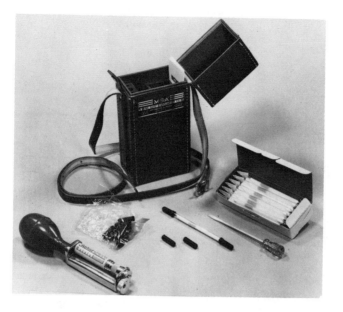

Fig. 7–19 Adsorbtion CO detector (Courtesy of Mine Safety Appliance Co.).

are placed in Dewar flasks, which contain in order from the air inlet: (1) ice and salt to maintain a temperature of $-16°C$, (2, 3, and 4) Dry Ice (CO_2), and acetone or Methyl Cellosolve to maintain temperatures of $-80°C$, and (5) liquid nitrogen to maintain a temperature of $-195°C$. Air is drawn through the traps by a vacuum pump (Fig. 7–21). After collection, the trap is allowed to warm up, and then the gases are analyzed by mass spectroscopy, infrared detection, or gas chromatography.

Ice crystals plugging the chambers are a major problem with the freeze-out condensation method. To avoid clogging, the inlet tube of the trap was flared to 2 in. in diameter and extended to within 2 in. of the bottom. The inlet and outlet tubes of the trap were placed into the traps and tubes designed with large radius bends to improve air-flow. Also, placement of drying towers on the inlet side of the trains will help reduce this problem. Keeping the bath solutions at a constant level and managing the bulky train are also problems.

Grab sampling is another technique for collecting gaseous pollutant samples. A *grab sample* is a sample taken at a particular time within an interval of a few seconds to a minute. A grab sample may also be a small representative portion removed from a gross sample

FREEZE-OUT EQUIPMENT FOR
ATMOSPHERIC SAMPLES

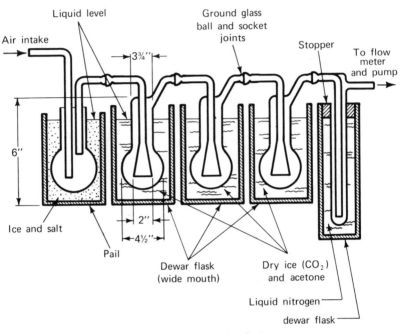

Horizontal Sampling Train

Fig. 7–20 Freeze-out sampling train (Courtesy of OMD-APCO-EPA).

without alteration. The sample is used to determine the gaseous composition of the carrier gas (air sample that contains the pollutant) and to determine the concentration of the pollutant at the instant the grab sample is collected.

Grab-sampling techniques are quite useful when the pollutant of concern is to be collected in a liquid and the rate of absorption is slow. Under these conditions, the absorbing solution is placed in the sampling apparatus, and the sample collected is allowed to come to equilibrium with the absorbing solution before analysis is made.

Most grab-sampling techniques utilize a minimum of equipment and require little special training or experience on the part of the operator.

Various types of grab samplers are available. One type is a *deflated plastic bag* (Fig. 7–22) placed in a closed box with a tube extended outside the box. A sample is collected by evacuating the air

Fig. 7–21 Vacuum pump (Courtesy of OMD-APCO-EPA).

from the grab sample box, thus causing the plastic bag to inflate, drawing the sample into the bag.

A *syringe* is essentially a solid displacement method sometimes used to collect small amounts of gas. Another device is an *evacuated flask* (Fig. 7–23) fitted with a stopcock or vacuum cap. In collecting the sample, a solution that efficiently absorbs the pollutant in question is added to the collection flask. The flask is then evacuated (at the site

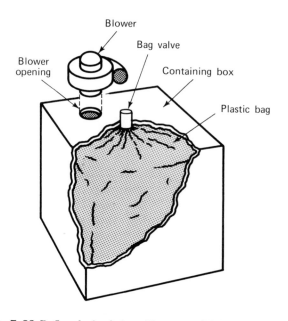

Fig. 7–22 Deflated plastic bag (Courtesy of OMD-APCO-EPA).

Fig. 7–23 Evacuated flask (Courtesy of OMD-APCO-EPA).

or in the laboratory) to the vapor pressure of the absorbing solution, the temperature and pressure of the flask are recorded, and the stop-cock closed. When the flask is opened, the sample is drawn into the flask. To reduce hazards from implosion (shattering of glass inward), these collectors should be placed in a protective container made of surethane foam or wrapped in adhesive or glass-fiber tape.

Another device used is a *purging device* or *gas-displacement collector* (Fig. 7–24), which is a cylindrical tube with stopcocks at both ends. When both stopcocks are opened, air is drawn into the collector; then both stopcocks are closed to contain the sample.

Another device used in grab sampling is the *liquid-displacement collector* (Fig. 7–25). Any suitable container in which liquid can be displaced may be used. As liquid is drained out of the container at the bottom, an air sample is sucked in at the top where the space is vacated by the liquid. The choice of liquid used (water, brine, mer-

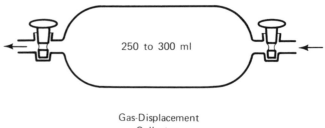

Gas-Displacement
Collector

Fig. 7–24 Gas displacement collector (Courtesy of OMD-APCO-EPA).

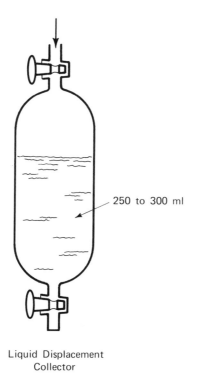

Liquid Displacement
Collector

Fig. 7-25 Liquid displacement collector (Courtesy of OMD-APCO-EPA).

cury or water saturated with the gas to be sampled) depends upon
the material being sampled. The gas being sampled must not react
with the liquid being displaced.

Grab sampling requires minimum equipment and minimum
training. It is preferable to continuous-monitoring sampling (covered
in the next section) when the constituents have slow absorption rates.
It is not profitable when small quantities of pollutants are present in
the sample, because measurable samples cannot be obtained.

Sampling devices, collection techniques, pollutants to be collected, and methods of analysis for gaseous sampling are summarized
in Table 7-2.

Continuous-monitoring (CM) equipment combining collection
with automatic analysis is available for routinely sampling many air
pollutants; however, these devices are used mostly for gaseous pollutants. The sampling train generally consists of a probe, absorber

TABLE 7-2   Gaseous Sampling Equipment

Sampling Device	Collection Technique	Pollutant to Be Collected	Method of Analysis
24-Hour Bubbler (Fig. 7-14) Sequential Sampler (Fig. 7-15)	Absorption with fritted glass absorbers (Fig. 7-16)	Inorganic gases, oxidants, NO_2, SO_2, H_2S Organic gases: acrolein, formaldehyde, aliphatic aldehydes, olefins	Wet chemistry, colorimetry, spectrophotometry
Detector tubes or indicator tubes (Figs. 7-17, 7-19)	Adsorption by activated charcoal, silica gel, activated alumina, molecular sieve	Insoluble or nonreactive vapors: NH_3, H_2S, SO_2, CO, NO_2, CO_2	Comparison with colors on a comparative chart
Lead peroxide candle (Fig. 7-1)	Absorption of SO_2 to form lead sulfate	SO_2	Measured in mg SO_2 per day/ 100 cm^2
Freeze-out sampling train (Fig. 7-20)	Condensation	Insoluble and nonreactive vapors, HC, radioactive gases	Mass spectroscopy, infrared or gas chromatography
Evacuated flask (Fig. 7-23), gas displacement collector (Fig. 7-24), liquid displacement collector (Fig. 7-25), plastic bag in a box (Fig. 7-22)	Grab sampling	Small quantities of gaseous pollutants—odor measurement	Varied

with a reagent, electronic detector amplifier, a meter, and a recorder. Some more recent instruments require no reagents, metering pumps, or fragile glass containers, and instead utilize solid-state electronic devices. Continuous-monitoring equipment is generally much more expensive than the sampling devices previously discussed; however, where heavy air pollution is common, a continuous-monitoring system may be an integral part of an air pollution warning system. In some large cities CM equipment is designed to set off an alarm system when a given pollutant reaches episode proportions. Pollutant sources are then required to reduce or stop operations until acceptable pollution levels are restored. Some pollutants for which CM equipment is designed include hydrocarbons, carbon monoxide, methane, sulfur dioxide, nitrous oxides, ozone, hydrogen sulfide, aldehyde, and total oxidant.

Hydrocarbon analysis has historically been achieved by total HC analyzers employing the flame ionization detector technique. This technique is preferred over nondispersive infrared (NDIR) because of its inherent sensitivity and capability of measuring methane from 0 to 1 ppm full-scale measurement and higher. Since the national ambient air quality standards now reflect "reactive HC (nonmethane portion)" as total HC less methane with reference to such an analysis method as chromatography, the *gas chromatograph*, such as Beckman's Model CG 6800 (Fig. 7–26), is more frequently used. This instrument monitors three air pollutants, CO, CH_4, and total HC from 0 to 1 ppm full scale for each component. This instrument has a precision digital timer to automatically control all functions when unattended. Front-panel controls and indicators facilitate calibration, monitoring, and operation.

With respect to CO determinations, the reference method defined in the national ambient air quality standards is NDIR. Beckman's Model 315 BL or Mine Safety Appliance Company LIRA (Fig. 7–27) are two examples of an NDIR analyzer for CO. A special equivalent method is gas chromatography with a flame ionization detector employing methanation of CO to CH_4. Here again Beckman's Model CG 6800 can be used for CO, as indicated previously.

Beckman has a series of *coulometric analyzers*: Model 906A (for SO_2), 908 (for total O_3), 909 (for NO), and 910 (for NO_2). These units are specifically designed for atmospheric (ambient) air monitoring. All units are basically similar in design and operate with minor differences in scrubbers and analysis system configurations. (Oxidant analyzer Model 908 is shown in Fig. 7–28.) These instruments have eliminated the need for reagents, metering pumps, fragile contactors, and other complex components.

Fig. 7–26 Gas chromatograph (Model 6800 courtesy of Beckman Instruments Inc.).

Technicon's Air Monitor IV (Fig. 7–29), which uses the *colorimetric technique,* has considerable flexibility in that with a minimum of downtime for servicing and maintenance and rearranging of components it can be programmed to measure SO_2, NO_2, NO_x, H_2S, F, aldehyde, or total oxidants. Table 7–3 lists some CM equipment.

ANALYTICAL PROCEDURES

Analytical procedures for interpreting particulate and gaseous pollutant data involve sampling, separation, concentration, development of a measurable property, measurement of a characteristic property, recording, calculating, and interpreting.

Fig. 7–27 Infra-red analyzer (Courtesy of LIRA-Mine Safety Appliance Co.).

sampling

Sampling has been discussed in previous paragraphs. This step often includes concentrating the sample.

separation

Separation removes substances that interfere with the development of measurable properties of the pollutant. For gaseous pollutants this can be done by chromatography, extraction, ion exchange, distillation, or crystallization. The only particulate analysis requiring separation is for a high-volume filter, when dealing with organic, inorganic, and radioactive compounds, or separation of particulates being observed microscopically for comparison with particle atlas pictures.

concentration

Concentration reduces the volume of a given substance. It is accomplished by pressurizing a gaseous pollutant to intensify the emission or absorption of radiation, or evaporating an inert solvent to produce a more concentrated solution for analysis. In some instances the concentrating step can be incorporated into the sampling or separating process. The high-volume filter combines sampling with concentration.

Fig. 7–28 Coulometric oxidants analyzer (Model 908 courtesy of Beckman Instruments Inc.).

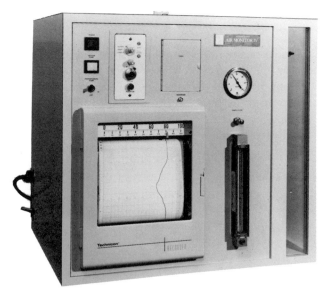

Fig. 7–29 Colorimetric analyzer (Courtesy of Air Monitor IV Technicon).

TABLE 7-3   CM Automatic Gaseous Pollutant Analyzers

Pollutant of Interest	Analytical Method	Manufacturer
Total hydrocarbons	Gas chromatography	Beckman Model 6800 (Fig. 7-26)
Carbon monoxide	Nondispersive infrared absorption	Beckman Model IR 315 BL
	Gas chromatography	Beckman Model 6800
	Nondispersive infrared absorption	Mine Safety Appliance Co. (Fig. 7-27)
Fluorides	Colorimetric	Technicon (Fig. 7-29)
Nitrogen dioxide	Coulometry	Beckman Model 910
	Colorimetric	Technicon
	Gas chromatographic	Varian Aerograph
Sulfur dioxide	Gas chromatographic	Varian Aerograph
	Flame photometry	Melpar
	Colorimetric	Technicon
	Coulometry	Beckman Model 906-A
Hydrogen sulfide	Colorimetric	Technicon
Aldehydes	Colorimetric	Technicon
Oxidants	Colorimetric	Technicon
	Coulometry	Beckman Model 908 (Fig. 7-28)

development of a measurable property

In many determinations it is necessary to develop some property that is sufficiently pronounced and distinctive to allow its measurement. One example is a process in which gases are brought into contact with specific chemicals and distinct color reactions evolve. These color intensities are easily measured. Gaseous materials that absorb ultraviolet (UV) or infrared (IR) wavelengths can be determined by spectroscopic procedures.

measurement of a characteristic property

Measuring is basic to the classification of many analytical procedures. Quantitative results can be obtained only by measuring the characteristic property or by comparing its intensity with some reference standard.

The *gravimetric method* isolates a substance or one of its compounds in the pure state and weighs it on an analytical balance. This process is tedious and lacks sensitivity. Some instruments using this technique are the high-volume filter, a dust bucket, impaction and impingement devices, a cyclone sampler, and a lead peroxide candle.

The *gasometric method* isolates a given substance in the gaseous form or is absorbed from gaseous mixture. The volume of the isolated pollutant is determined by comparing the weight of the mixture before and after the gas is removed, or a gas buret (graduated glass tube used to measure quantities of gas received or discharged) is used to determine the volume of gas removed. This method is used in stack sampling but not in atmospheric sampling, because it lacks sensitivity.

The *titrimetric method* utilizes a liquid buret that releases a titrant (a solution containing a known concentration of reactive chemical) drop by drop into an air sample. This produces a color reaction, an electrical potential reaction, or a photometric reaction or end point. The amount of titrant required to attain an end point is a measure of the amount of the component being measured in the sample.

The *absorption of radiation method* permits both identification and quantitative estimation of gases. Various light wavelengths, including UV and IR rays, are utilized in this method.

Atomic absorption spectroscopy is one absorption of radiation method used on heavy metals. The element of interest is heated to a high temperature in a flame until chemical bonds between molecules are broken and the atoms are allowed to float freely in an atomic vapor. In this condition the atoms are not excited (i.e., orbital electrons are not raised to higher energy levels), but merely are dissociated from molecular bonds and placed in an unexcited, normal, or ground state where the element of interest is capable of absorbing radiation at specific wavelengths. It has been found that light energy from a cathode of the appropriate wavelength can be projected into the atomic vapor and will be absorbed. The amount of absorption can be monitored by a light detector, and this measurement is related to the amount of the element present in the sample.

Colorimetry (refers to visible regions of the light spectrum) is another absorption of radiation method used to detect transition metals as they form highly colored salts when reacting with certain reagents. Inorganic ions (sulfite, nitrite, fluoride, metal ions) and many organic compounds, which are themselves colorless, will also produce a color when a carefully chosen reagent is added. This color can be compared against a standard chart, a photoelectric colorimeter (filter photometer), or a spectrophotometer. A glass prism or diffraction grating in a spectrometer replaces the filter in a photoelectric color-imeter to produce monochromatic (one-color) light.

A *transmissometer* or *densitometer* used with a paper tape sampler is another absorption-technique application. The percentage of light transmitted and/or the optical density shows up on the tape (Fig. 7–5). A coefficient of haze (COH) per 1,000 linear feet is computed as follows:

$$COH/1,000 \text{ linear ft} = \frac{(\text{optical density}) (10^5) (A)}{V}$$

where A = area of sample spot in square feet and V = volume of air sampled in cubic feet. Optical density is read from a densitometer. Calibrate the unit to maintain 15 standard ft^3/h (SCFH).

The soiling index rating system is as follows:

COH/1,000 Linear Feet	*Rating*
0.0 – 0.9	Light
1.0 – 1.9	Medium
2.0 – 2.9	Heavy
3.0 – 3.9	Very heavy

The **emission of radiation method** is the basis for several approaches for ascertaining pollution concentrations, especially of metallic elements.

The principle of *emission spectroscopy*, like atomic absorption spectroscopy, heats an element of interest to a high temperature with a flame until chemical bonds between molecules are broken and atoms are allowed to float freely in an atomic vapor. However, by application of more energy, the normal or ground state of the electrons is raised to a higher energy or *excited state*. Since this higher energy level is unstable, the electrons will return to their original ground state, accompanied by emissions of radiant energy that are measurable. The intensity of characteristic wavelength emitted indicates the substance or element implicated.

Flame photometry is based on the excitement of a pollutant with a flame and then measurement of the light emission by photographic means.

Fluorimetry is based on exciting a sample with a light (usually UV) and measuring the intensity of the resulting fluorescence as a measure of pollutant concentration.

Two radiation-emission approaches quantify the solid materials present in colloidal suspensions (suspension of particles too small for resolution with an ordinary microscope). *Turbidimetry* measures white-light transmission through a finely divided suspension and then compares it with a standard suspension. *Nephelometry* compares white light reflected from within the solution containing the suspension.

The **electrical method** has many ways for measuring characteristic properties. *Conductimetry* measures a solution's ability to carry an electric current between two electrodes. *Polarography* introduces suitable electrodes into a solution, a small voltage is applied, and the amount of current flow is determined. (*Amperometric* is a special adaptation of the polarography technique in which current is measured in amperes.) *Coulometry* measures the equivalent relationship between the quantity of electricity and quantity of chemical change. *Flame ionization* stimulates the air sample with a flame, and the ionization current is measured as electrons and positive ions are attracted to a polarized electrode. *Potentiometry* uses the electrical potential or voltage across two electrodes as an indication of the ion concentration in the solution.

The **microscopic method** can be used to size and identify particulates under the microscope which have been extracted from a high-volume filter, from a cascade impactor, or from an electrostatic precipitator.

The **counting method** may be used on spores, pollen, and bacteria. They are either counted under a microscope or they are counted directly. Particles collected on sticky tape can be compared to the Gruber particle counter (Fig. 7–30).

Chromatography is a method of analysis in which a flow of solvent or gas promotes the separation of substances by differential migra-

Fig. 7–30 Gruber particle compatator (Courtesy of Research Appliance Co.).

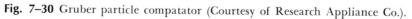

tion from a narrow initial zone in a porous sorptive (adsorptive or absorptive) medium. *Column chromatography* is based on selective adsorption of the substances on a solid material. *Partition chromatography* is based on substances being separated to various degrees between two immiscible (incapable of mixing) liquids. *Paper chromatography* is an example of this. *Ion-exchange chromatography* involves an exchange of ions between a solid and a solution. Column chromatography is one method used in analyzing polynuclear hydrocarbon particulates collected on a high-volume sampler. *Thin-layer chromatography* is performed on thin layers of adsorbent materials that have been spread on a suitable support. This method is used for analysis of many organic gaseous and particulate pollutants. *Gas chromatography* is a method for separating the components of a mixture of compounds of sufficient volatility to be vaporized. This method is used in such instruments as Beckman's Model CG 6800 for detection of total hydrocarbons.

recording

Standard record forms should be established for recording manually obtained data. Strip-chart or digital recorders offer many advantages in many instances. Some recording systems are keyed to computers that warn of impending air pollution episodes by tripping an alarm. Computerized data allows speedy retrieval of all recorded data.

Comparing air pollution data is difficult because many recording systems have been used. To ensure correlation in the future, standard units for expressing these data have been recommended (Table 7–4). In the past, gaseous pollutants have been recorded in parts per million

TABLE 7-4   Recommended Units for Expressing AP Data

Particle fallout	Milligrams per square centimeter per time interval $(mg/Cm^2/MO$ or $mg/Cm^2/yr)$
Airborne particulates	Micrograms per cubic meter $(\mu g/m^3)$
Particulate counting	Number per cubic meter
Gaseous pollutants	$\mu g/m^3$ (except for CO expressed as mg/m^3)
Gas volumes	Reported as standard. Corrected to $25°C$ and 760 mm mercury pressure
Temperature	Centigrade—$°C$
Time	Twenty-four hundred hour clock (0000 to 2400 hours)
Pressure	Millimeters of mercury (mm of Hg)
Linear velocity	Meters per second (m/sec)
Volume emission rates	Cubic meters per minute (m^3/min)
Sampling rates	Cubic meters per minute (m^3/min)
Instantaneous light transmission	Percent transmittance
Visibility	Kilometers—km

(ppm). Since different gases have varying physical characteristics, specific conversion factors are required to convert parts per million to the presently accepted micrograms per cubic meter $(\mu g/m^3)$. Conversion factors for the gaseous pollutants of major concern are listed in Table 7–5.

calculating

Calculating is necessary to determine the weight or volume of the air sampled and weight or volume of the pollutants within the sample. The pollutant concentration is then expressed in specific terms

TABLE 7-5   Conversion Factors

O_3	1 ppm	=	$1,960\ \mu g/m^3$
	$1\ \mu g/m^3$	=	0.51×10^{-3} ppm
SO_2	1 ppm	=	$2,860\ \mu g/m^3$
	$1\ \mu g/m^3$	=	3.5×10^{-4} ppm
CO	1 ppm	=	$1.15\ mg/m^3$
	$1\ mg/m^3$	=	0.87 ppm
HC	1 ppm	=	$655\ \mu g/m^3$
	$1\ \mu g/m^3$	=	1.53×10^{-3} ppm
NO_2	1 ppm	=	$1,880\ \mu g/m^3$
	$1\ \mu g/m^3$	=	0.53×10^{-3} ppm

(Table 7–4). Results are sometimes expressed as an *arithmetic mean*—the sum of a given number of factors divided by the number of factors (e.g., $3 + 4 + 5 + 6 = 18 \div 4 = 4.5$). Sometimes results are expressed as a *geometric mean*–the Nth root of the product of N factors (N = number of factors) (e.g., $3 \times 4 \times 5 \times 6 = \sqrt[4]{360} = 4.35+$).

interpreting

Interpreting is important when data must be normalized (adjusting values by some factor) to ensure that the summation of values will not exceed 100 per cent. To ease interpretation, large volumes of data are normally projected on probability graphs, frequency-distribution curves, bar graphs, or tables.

REFERENCES

Magill, P. L., F. R. Holden, and Charles Ackley, ed., *Air Pollution Handbook*. New York: McGraw-Hill Book Company, 1956.

Stern, A. C., ed., *Air Pollution*, Vol. I. New York: Academic Press, Inc., 1962.

U.S. Environmental Protection Agency, *Federal Register*, National Ambient AQ Standards,

Vol. 36, No. 21, pp. 1502–1515, Jan. 30, 1971.
Vol. 36, No. 67, pp. 6680–6701, Apr. 7, 1971.
Vol. 36, No. 84, pp. 8186–8201, Apr. 30, 1971.
Washington, D.C.: U.S. Government Printing Office.

RECOMMENDED FILMS

TSA-37   Measurement of Particulate Pollutants (35-mm slide set)
MA-43   Tape Sampler Calculations (20 min)
TF-112   Collection of Particulate Matter
  Available: Distribution Branch
        National Audio-Visual Center (GSA)
        Washington, D.C. 20409

QUESTIONS

1/ What is the difference between atmospheric air sampling and source sampling?

2/ What are two purposes of atmospheric air sampling?

3/ Explain why size, rate and duration of sampling are important factors to be considered in determining equipment necessary for sampling.

4/ What device is used to collect the following pollutants; what technique of collection is involved; what method of analysis is used; and in what units are results expressed? Settleable particulates, suspended particulates, pollen, bacteria, NO_2, HC, NH_3, odor-producing pollutants.

5/ Explain the difference between absorption and adsorption.

6/ Name several grab-sampling devices.

7/ Explain the difference between colorimetry and coulometry.

8/ Explain the difference between atomic absorption and emission spectroscopy.

9/ What is the difference between turbidimetry and nephelometry?

10/ What are normalized data?

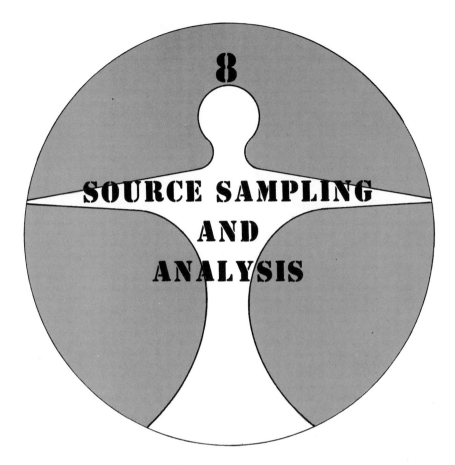

8

SOURCE SAMPLING
AND
ANALYSIS

Student Objectives

—*To develop an understanding of source sampling.*
—*To become familiar with stack sampling train components
for particulate sampling.*
—*To become aware of methods used for gaseous source sampling.*
—*To learn how to evaluate smokestack plumes and plumes
from mobile sources.*

Source sampling and analysis are performed to determine the
amount of air pollution *emitted from a specific source*, as differing
from atmospheric sampling and analysis, which deals with pollutants
within the total air mass surrounding the earth.

Source sampling includes *stack* sampling performed by inserting a measuring device into the smokestack of a particular source to obtain a sample, or *plume evaluation* or "smoke reading" of visual emissions from a *stationary* source. Source sampling also includes insertion of a probe into the exhaust to extract a sample for analysis, or plume evaluation of exhaust emission from a *mobile* source.

PURPOSE OF SOURCE SAMPLING

The purpose of source sampling is to determine the quantity and type of pollutants emitted from a specific source to determine compliance or noncompliance with an emission standard, to determine the efficiency of a pollutant collector or control device, to determine emission factors for use in emission inventories, or to determine appropriate design for air pollution control equipment to be installed.

STACK-SAMPLING PRINCIPLES

The collection method and collection equipment used for stack sampling depend upon the purpose of the sampling, but also hinge upon the physical environment, for example, very high temperature operations, or extremely moist, dusty, or corrosive environments. The accessibility of sampling locations can also dictate the type of sampling equipment. For example, heavy bulky equipment may not be possible to use if one is required to scale a difficult stack to obtain an effluent sample. It is important to remember that the sample must be collected without physical or chemical alteration and without altering the flow pattern or the concentration of pollutant at the point of collection of the sample. Also, this sample must be obtained at a point of average gas density and average pollutant concentration.

STACK SAMPLING AND ANALYSIS FOR PARTICULATES

The *primary* purpose of particulate stack sampling is to determine the weight of solids, and sometimes liquids and condensable vapors, in a specific volume of stack gas. Therefore, the velocity of the stack gas or carrier gas must be measured if the particulate emission rate is to be determined.

There are five major components of the equipment used to obtain the particulate sample. These components collectively are referred to as the *sampling train*. The train includes a pitot tube (pitobe), the sample box, the duorail, the umbilical cord, and the meter box (Fig. 8–1).

pitobe

The pitobe is a combination of a sampling probe for insertion into the stack and the Pitot tube that is used with a manometer to measure velocity of gaseous flow in the stack. In sampling for particulates the sample must be a representative *isokinetic* sample. Therefore, the pitobe nozzle must be inserted into the stack at a point where the velocity measured through the collecting nozzle will be equal to the effluent velocity in the stack.

The pitobe is used to make a preliminary traverse of a stack or a duct to determine an isokinetic flow point. It is inserted into the sampling port for 5 or 10 min at each selected traverse point to allow a differential or velocity pressure reading. With rectangular ducts, the traverse is made by dividing the duct into equal areas and sampling at the center of each area. With circular ducts, equal concentric areas are sampled in the center of each area on both sides of the center. The number of equal areas chosen depends on the size of the duct and the accuracy desired. Round sections are preferable to square or rectangu-

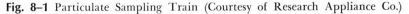

Fig. 8–1 Particulate Sampling Train (Courtesy of Research Appliance Co.)

lar sections because of less turbulence or variation in gaseous flow than often occurs in sections with corners. The ideal sampling location in a duct should be at least seven diameters downstream and three diameters upstream from any point of flow turbulence in order to obtain more uniform gaseous flow rates. A branch entry, an elbow, a damper, or an outlet of a fan or collector are all suggested locations.

sample box

The sample box is assembled with a glass cyclone that removes particles greater than 5 μ in diameter, and, following the glass cyclone, a glass holder with a coarse-porosity fritted glass filter for smaller particles. Both cyclone and filter are maintained between 240 and 280°F in a heated portion of the sample box to ensure that the gas remains above its dew point to prevent condensation of water vapor at the filter. Next in line are four Greenburg–Smith impingers placed in an ice bath to force condensation of water vapor from the gas stream. Only the second impinger has the original impinger tip; the others have the tip removed to decrease pressure drop through the train. The first impinger contains 250 milliliters (ml) of deionized water, the second, 150 ml of deionized water. The third impinger is left dry to remove entrained water; the fourth impinger contains 175 g of silica gel to remove any remaining water. In some trains these impingers may be used as wet collectors to capture particles. Electrical circuitry provides heat to the heated compartment, which is separated from the ice bath by polyethylene foam insulation.

duorail

The duorail supports the pitobe and the sample box on the stack. It is attached to two aluminum rails on the sides of the box, which allow either horizontal or vertical movement. An angled aluminum brace fastened to the stack by a nylon cord supports the rails. In other words, the duorail is one leg of a right triangle, the nylon cord (paralleling the stack) is the second, and the brace forms the hypotenuse. This device fastens the sampling equipment to the stack at the sampling port.

umbilical cord

The umbilical cord connects the last modified Greenburg–Smith impinger, the Pitot tube, and the heating elements to the meter box.

meter box

The meter box contains the vacuum pump, regulating valves, volumetric flow meters, Pitot tube manometers, vacuum gauge, and electrical controls. A calibrated orifice and an inclined vertical water manometer indicate the sample rate. A dry gas meter and stopwatch are used to determine the integrated gas-sample volume.

The pitobe is attached to the sample box, which is located in a convenient place within 100 ft of the sample port. The umbilical cord is connected to the meter box and the sample box. The pitobe is usually inserted into the sampling port for 5 or 10 min at each selected traverse point, and a differential or velocity pressure reading is obtained from the pitobe manometer. The sample rate is maintained isokinetically by observing the pitobe manometer reading, performing computations interpolated from a nomograph, and correcting the sampling rate so that pressure drop across the calibrated orifice corresponds to stack velocity.

accessories

1/ A *U-tube manometer* is inserted into the stack to determine static duct pressure.

2/ A *barometer* is used to record atmospheric pressure.

3/ A *wet–dry bulb thermometer* is inserted to obtain data to compute moisture content of the carrier gas.

4/ A *dial thermometer* is inserted to obtain stack gas temperature.

5/ An *ORSAT* analyzer* measures CO, CO_2, and O_2 content of the carrier gas, which is used to determine the dry molecular weight of the carrier gas.

These data in turn are used to determine the velocity of the carrier gas. Computation of the area of the cross section of the stack being sampled, coupled with velocity data, gives the volume of the carrier gas.

An inexpensive although time-consuming analysis of stack gases can be performed by a mechanical ORSAT produced by Burrell Corporation (Fig. 8–2). This device utilizes absorption and oxidation principles and can analyze stack gases for CO, CO_2, O_2, and several HC vapors. The acid used in mechanical ORSAT's makes them hazardous, and they need refined technical maintenance.

*ORSAT, oxidation–reduction selective absorption technique.

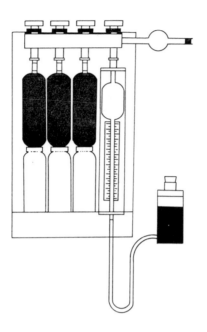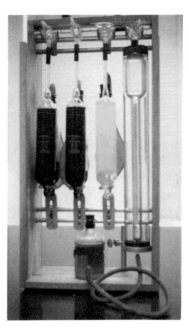

Fig. 8–2 ORSAT Analyzer (Courtesy of Burrell Corp.)

Harvey-Westbury Corporation has produced an electronic ORSAT (Fig. 8–3) that eliminates tedious time-consuming analytical processes. Using gas chromatography separation, the electronic ORSAT can measure N_2 in addition to all the other gases that the mechanical ORSAT can analyze. Any recorder may be accommodated to the electronic ORSAT. This electronic ORSAT and recorder are more expensive than the mechanical equipment, but provide highly accurate data.

Another highly accurate device, produced by Mine Safety Appliance Company (Fig. 7–27), utilizes an infrared principle for analysis of CO, CO_2, and various HC vapors in flue gas. Although more expensive than the mechanical ORSAT, this device also eliminates time-consuming analysis of stack gases. It also provides a means of investigating automobile exhaust gases as well as stack gases.

Depending upon the carrier gas being sampled, various types of filters may be attached near the Pitot tube to collect particulates. Ceramic filters made of Alundum placed in thimbles are used where it is important to have more strength when encountering wet collection conditions, chemical resistance, or high temperature resistance. Paper thimbles or discs are accurate, inexpensive, and convenient

Fig. 8–3 ORSAT Analyzer (Courtesy of Harvey-Westbury Corp.)

filters to use. Glass-cloth filter discs or thimbles are often used for coarse dust, followed by paper thimbles for fine dust. Extremely small particles may be collected with cellulose ester or membrane filters. Particulates collected on filters are weighed and expressed as weight per standard cubic meter or as weight per unit of time. When impingers containing liquids are used to collect particulates, the solid sample is obtained by evaporating the liquid.

STACK SAMPLING AND ANALYSIS OF GASEOUS POLLUTANTS

The purpose of gaseous pollutant stack sampling is to determine the gaseous pollutants contained in a carrier gas stream emitted from a stack. The components of the *carrier gas* stream include the *pollutant gases*; however, all components of the carrier gas stream are not considered as pollutants. Just as in particulate stack sampling, the velocity of the carrier gas stream must be measured if the pollutant emission rate is to be determined.

grab-sampling methods

The grab-sampling methods used in atmospheric sampling and discussed in Chapter 7 are also used for stack sampling for gaseous pollutants. These methods make use of a plastic inflatable bag, liquid-displacement collector, gas-displacement collector, evacuated container, and syringe.

A plastic inflatable bag enclosed in a grab-sample box is most often used to collect pollutant gases. The bag is conditioned by pumping in the stack gas to be sampled two or three times and expelling it before an actual sample is taken for analysis to ensure no residual of gas remains other than the stack gas. A sample is collected by evacuating the air from the grab-sample box, causing the flexible bag to inflate and drawing the sample out of the duct. The gas is expelled from the bag for analysis by reversing the flow of air from the pump used to evacuate the air from the box when the sample is collected.

If a liquid-displacement collector is used, it should contain a solution that absorbs the pollutant to be collected and will not dissolve or react with the gas sample components.

If an evacuated container is used with an absorbing solution, the flask is evacuated until it reaches the vapor pressure of the absorbing solution, the temperature and pressure of the flask are recorded, the clamp on the tubing is tightened, and the glass plug is placed in position. After clearing residual air from the probe to be inserted into the stack, the flask is opened to draw the sample out of the stack into the flask, and the flask is closed and returned to the lab for analysis.

Grab sampling can be used to collect a sample for the ORSAT, wet chemistry, infrared analysis methods, or gas chromatography.

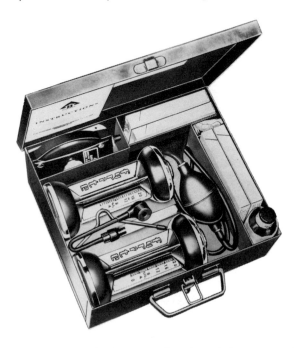

Fig. 8–4 Flue Gas Kit (Courtesy of Bacharach Instruments Co.)

Fig. 8–5 Carbon Monoxide Indicator (Courtesy of Bacharach Instrument Co.)
Fig. 8–6 Carbon Dioxide Indicator (Courtesy of Bacharach Instrument Co.)

Another grab-sampling method that combines sampling with analysis utilizes a flue gas kit produced by Bacharach Industrial Instrument Corporation (Fig. 8–4), which may be used in lieu of the ORSAT analyzer described earlier. This kit contains a carbon monoxide indicator (Fig. 8–5), which contains a yellow-colored chemical gel that forms a darkish brown stain on contact with CO. The percentage of carbon monoxide or parts per million is read directly by comparing the length of the stain in the tube to an etched metal scale. A Fyrite oxygen indicator and a carbon dioxide indicator (Fig. 8–6) are also included in the kit. From these oxygen and carbon dioxide can be read in percentages. This kit employs the fundamental method of volumetric measurement involving chemical absorption of a 60-cc gas sample by a chemical absorber.

When large sample volumes or average concentrations are desired, continuous sampling with a sampling train is preferable to grab sampling, since grab samples, due to their small size, are less statistically accurate. The stack-sampling train used for gaseous pollutants is similar to the one used for particulates, except that the filters are removed if they interfere in the analysis, and the pollutant collectors are absorbers, adsorbers, or freeze-out traps. When sampling for gaseous pollutants, the sample must be *proportional*. That is, the velocity at the tip of the probe must be proportional to the velocity of the approaching gas stream.

sampling techniques for specific gaseous pollutants

Sulfur dioxide in stack gases can be determined by a colorimetric technique (West–Gaeke method), and acidimetric technique (Berk–Burdick method), or a titration technique (IPA–Thorin method).

Nitrogen oxides are identified by a colorimetric technique (phenoldisulfonic acid method). The absorbent consists of hydrogen peroxide in diluted sulfuric acid, oxidized to form nitric acid. The nitric acid nitrates phenoldisulfonic acid, which in turn forms a yellow compound when reacting with ammonium hydroxide. The color intensity produced is proportional to the concentration of NO_x in the sample. A colorimeter is used to measure this color intensity.

Fluoride is most often determined by the specific ion electrode method (Fig. 8–7), although SPADNS* zirconium lake method has been used.

Chlorine in stack samples is analyzed by using a modified Volhard titration method, the orthotolidene method, or a specific ion electrode method.

PLUME EVALUATION OF SMOKESTACK EMISSIONS

Ringelmann chart

The Ringelmann chart and the smokescope are two ways to measure air pollutants after emission from smoke stacks or after emission from mobile sources. Plume evaluation of mobile sources will be covered later. The Ringelmann chart was one of the first aids devised to visibly measure air pollutant emissions. It was developed in 1897 by a French engineer, Maximillian Ringelmann. Initially, it was used to assess the efficiency of fuel combustion, but it has become a legal device for determining whether smoke emissions are within the limits established by emission standards. The chart is made up of graduated shades of gray, which vary by five equal steps between white and black.
Information Circular 8333, *Bureau of Mines Smoke Chart*, contains four grills representative of the shades 1 through 4 of the Ringelmann chart. Shade 5 has no representative grill since it is 100 percent

*SPADNS—4, 5 dihydroxy —3 (p—sulfophenyl 220)—2, 7—napthalene disulfonic acid, trisodium salt or Sodium 2—(parasulfophenyl 220—1, 8—dihydroxy—3, 6—napthalene disulfonate.

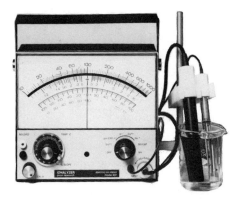

Fig. 8–7 Specific Ion Analyzer (Courtesy of Orion Research)

black. A miniature smoke scale card (Fig. 8–8) is available that provides an open aperture through which the smoke emission color can be compared to a color on the chart.

Smoke readers are usually trained at a visible emissions school lasting 1 to 3 days. The school has a smoke generator (Fig. 8–9), which is a mobile or stationary apparatus constructed to produce different degrees of black smoke from a simulated outdoor stack when benzene is burned in a combustion chamber. The density is varied by adjusting fuel injection rates, and a transmissometer mounted on the stack measures light transmission through the smoke. The apparatus is calibrated with several grades of neutral density filters in order to determine what settings will produce the varying shades of gray comparable to the Ringelmann chart. These settings can then be compared with students' visual sightings to grade their efficiency in evaluating the plume density.

The smoke generator also produces white smoke by vaporizing number 2 fuel oil in the exhaust manifold of a gasoline engine. This nonblack smoke is evaluated on an equivalent opacity or degree of transmitted light obscurity. Equivalent opacity is related to the Ringelmann chart numbers as follows:

Black Smoke Intensity	Nonblack Smoke Opacity (%)
1	20
2	40
3	60
4	80
5	100

(In other words, black smoke intensity of Ringelmann no. 1 is comparable to a nonblack smoke intensity of 20 percent opacity.) Number 1 smoke is used as a standard, and the percentage density of the smoke is obtained by the formula

$$\frac{\text{equivalent units of no. 1 smoke} \times 20\% \times 100}{\text{'number of observations'}} = \% \text{ smoke density}$$

The smoke plume can be read with an accuracy of one fourth intensity or 5 percent opacity (e.g., readings of black smoke may be 1.25 and readings of white smoke may be 25 percent).

To provide most accuracy in reading smoke, there are certain conditions prescribed for the location of the observer and his relationship to the light source and the plume. The observer should be positioned not closer than 100 ft and not further away than $\frac{1}{4}$ mile from the stack producing the smoke plume. He should be viewing the smoke plume at right angles to the smoke's direction of travel. The sun should be at the observer's back to provide light for viewing the plume without causing a glare in his eyes. The background immediately beyond the top of the stack should be free of buildings or other dark objects that could prevent clear visibility of the plume. Observations are repeated at $\frac{1}{4}$- or $\frac{1}{2}$-min intervals to prevent eye strain that could lead to poor evaluation from continuous viewing.

To qualify as an expert smoke reader, a student's average deviation must not be more than 10 percent and no reading may vary by one Ringelmann (20 percent opacity) or more for a set of 50 consecutive readings.

smokescope

The smokescope, developed by the Mine Safety Appliance Company, was designed to overcome some of the disadvantages of the Ringelmann chart, such as (1) variations in the background against which the smoke is viewed, (2) variations in the ambient light that illuminates the chart, which may be different from light in the area of the stack, and (3) limitations of the human eye to refocus in glancing from the smoke to the chart.

In using the smokescope, the observer views the stack through an instrument, aiming it so that smoke fills the field of vision through apertures. Light from an area adjacent to the stack is transmitted through a film disc in the barrel to the surface of a mirror. From this mirror an image of the reference disc (film disc in the barrel) is pro-

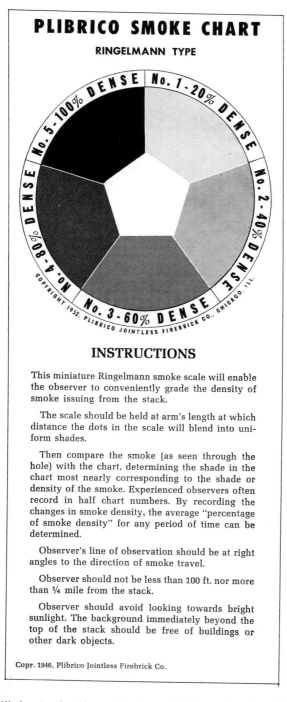

PLIBRICO SMOKE CHART
RINGELMANN TYPE

NO. 5 - 100% DENSE · NO. 1 - 20% DENSE · NO. 2 - 40% DENSE · NO. 3 - 60% DENSE · NO. 4 - 80% DENSE · COPYRIGHT 1952. PLIBRICO JOINTLESS FIREBRICK CO., CHICAGO. ILL.

INSTRUCTIONS

This miniature Ringelmann smoke scale will enable the observer to conveniently grade the density of smoke issuing from the stack.

The scale should be held at arm's length at which distance the dots in the scale will blend into uniform shades.

Then compare the smoke (as seen through the hole) with the chart, determining the shade in the chart most nearly corresponding to the shade or density of the smoke. Experienced observers often record in half chart numbers. By recording the changes in smoke density, the average "percentage of smoke density" for any period of time can be determined.

Observer's line of observation should be at right angles to the direction of smoke travel.

Observer should not be less than 100 ft. nor more than ¼ mile from the stack.

Observer should avoid looking towards bright sunlight. The background immediately beyond the top of the stack should be free of buildings or other dark objects.

Fig. 8–8 Plibrico Smoke Chart (Courtesy of Plibrico Jointless Firebrick Co.)

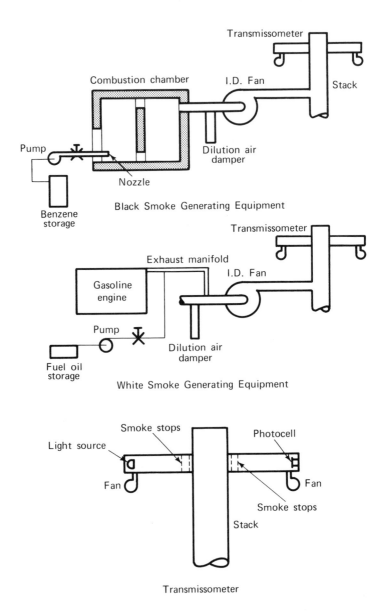

Fig. 8–9 Smoke Generator (Courtesy of OMD-APCO-EPA).

Fig. 8–10 Exhaust Emission Tester (Courtesy of Sun Electric Co.)

Fig. 8–11 Hydrocarbon Analyzer, Model 400 (Courtesy of Beckman Instruments Inc.)

jected onto the image mirror of the aperture, where it may be compared with the observed smoke without refocusing the eye. The main disadvantage of the smokescope is its limitation to only two readings on the Ringelmann scale, since one half the density disc is equivalent to no. 2 Ringelmann and the other half is equivalent to no. 3 Ringelmann.

SOURCE SAMPLING OF MOBILE SOURCE EMISSIONS

Air pollution from mobile sources is covered in Chapter 11. The Clean Air Act establishes the right of the federal government to set emission standards for all *new* vehicles and to test these vehicles to ensure compliance. Table 11–1 lists the standards established since 1968.

The federal government conducts a source-sampling exhaust emissions test to determine HC, CO, and NO_x mass emissions from new automobile exhausts. A flame ionization detector is used to measure HC, a nondispersive infrared detector is used to measure CO, and a chemiluminescence test is used to measure NO_x.

The state governments are responsible for testing vehicles *after sale* to the ultimate purchaser. Some states have established emission standards for all mobile sources. Old-model cars, which may have no control devices (produced prior to 1961–1962), may also be tested if a

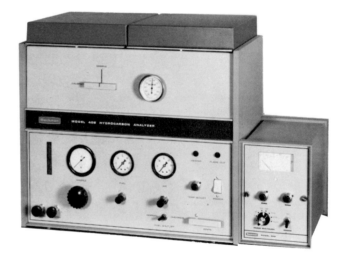

Fig. 8–12 Hydrocarbon Analyzer, Model 402 (Courtesy of Beckman Instruments Inc.)

state desires to do so. A few states use the Sun Electric Corporation exhaust emission tester (Fig. 8–10), which utilizes infrared analyses to measure HC in parts per million and CO in percentage. The pollutant sample is obtained from the vehicular source by inserting a probe into the tailpipe to extract exhaust flow for passage through the analyzer. The Beckman Model 400 hydrocarbon analyzer (Fig. 8–11) is used primarily for monitoring automobile exhausts and the Model 402 (Fig. 8–12) is designed for direct monitoring of exhaust from automobile diesels and jet engines.

Plume evaluation by the Ringelmann method is also used to evaluate visible emissions from gasoline- and diesel-powered vehicles. Some states have established a limit of Ringelmann no. 1 as acceptable for both gasoline and diesel vehicles.

REFERENCES

Berk, A. A., and L. R. Burdick, *A Method of Test for SO₂ and SO₃ in Flue Gases.* Washington, D.C.: Bureau of Mines Report of Investigations 4618, Jan. 1950.

Determination of SO₂ and SO₃ in Stack Gases, Emeryville Method Series 4516/59a, Shell Development Co. Analytical Dept. 50 W 50th, New York, N.Y. 10020, 1959.

Stack and Duct Sampling for Gases and Particulates, Tech. Bull. No. 7. Ann Arbor, Mich.: Gelman Instrument Co., n.d.

Standard Method of Test for Oxides of Nitrogen (Phenoldisulfonic Acid Produce), *Book of ASTM Standards,* ASTM Designation D-1608-60, pp. 725–729, 1965.

U.S. Department of Health, Education, and Welfare, *Optical Properties and Visual Effects of Smoke Stack Plumes,* PHS, Pub. No. 999-AP-30. Washington, D.C.: U.S. Government Printing Office, 1967.

U.S. Department of Interior, *Bureau of Mines Smoke Chart,* Information Circular No. 8333. Washington, D.C.: U.S. Government Printing Office, May 1967.

West, P. W., and G. C. Gaeke, Fixation of SO₂ as Disulfitomercurate (II) and Subsequent Colorimetric Estimation, *Analytical Chemistry,* Vol. 28, pp. 1816–1819, 1956.

RECOMMENDED FILMS

OE-367   Flue Gas Analysis (ORSAT apparatus) (19 min)

MA-48c (or MA-60)   Reading Visible Emissions (3 to 4 min or 35-mm slide series)

MA-51   Source Sampling Equipment (10 min)
   Available: Distribution Branch
             National Audio-Visual Center (GSA)
             Washington, D.C. 20409

QUESTIONS

1/ What are the two basic means of source sampling to determine stack emissions?

2/ What is meant by isokinetic sampling and when is it required?

3/ What is meant by proportional sampling and when is it required?

4/ What are the five major components of a source sampling train when stack sampling for particulates?

5/ What is one device that may be used to collect stack samples of gaseous pollutants?

6/ What is the purpose of the ORSAT analyzer? What alternative equipment may be used for determining the same gaseous components?

7/ What is the system most often used to measure visible air pollutant emissions?

8/ How is black smoke intensity rated?

9/ How is nonblack smoke intensity rated?

10/ How is source sampling accomplished with mobile sources?

11/ What are the major air pollutants from automobiles?

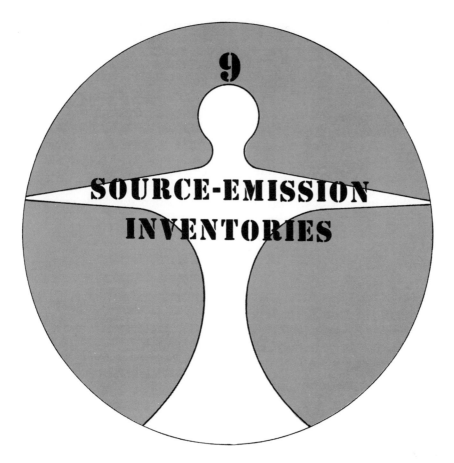

9

SOURCE-EMISSION INVENTORIES

Student Objectives

—*To understand the purpose of a source–emission inventory.*
—*To learn where to obtain data required to complete a source–emission inventory.*
—*To learn how to compute fuel consumption for stationary combustion sources and mobile combustion sources.*
—*To learn how to determine emission factors.*
—*To understand how the results of a source–emission inventory are presented and data interpreted.*

A *source inventory*, as defined in Chapter 2, provides a means of listing air pollution sources according to quantity and quality of materials processed. For convenience they may be classified as specific sources (Table 2–1) or multiple sources (Table 2–2).

An *emission inventory* involves an investigation of the sources within a given area to determine the amount of pollutants of various types being emitted from each source.

These two types of inventories may be considered as one procedure and be referred to as a *source–emission inventory*. A source–emission inventory includes the location, enumeration, and tabulation of pollution sources within a defined area, along with the type and quantity of pollutants being emitted from those sources. Data obtained from this inventory are essential for program planning, to include establishing control strategy, developing emission standards and control regulations, and preventing future pollution. This inventory also provides data needed for establishing permit systems for routine enforcement and for public information and community relations presentations.

A comprehensive source–emission inventory includes field visits, plant surveys, questionnaires, and stack sampling. When budgetary and personnel resources are limited, a rapid survey technique as described in Public Health Service, Publication No. 999-AP-29, may be utilized to provide reasonable working estimates.

INITIAL PREPARATION

The specific pollutants of consequence in the area should be determined and listed as the pollutants to be reported in the inventory. A decision must be made on whether to perform a comprehensive survey or a rapid survey inventory.

The geographical area to be surveyed must be defined. A map showing natural features, major arterial streets, and political boundaries and having a superimposed universal mercator grid is desirable. Since much data are tabulated by census tracts, these census tracts should be delineated. Reporting zones (groupings of census tracts) based on land use will usually delineate residential, commercial, or industrial areas. Residential areas may be further divided into subgroups based on population density. To assist in definition of differences in fuel types used, it may be helpful to separate residential areas into single-family-unit areas and apartment areas. Reporting zones should be approximately 2 to 10 square miles in area.

COLLECTION AND USE OF PRIMARY INFORMATION

Air pollution data may be tabulated as *area wide* or background source information, such as residential, commercial, governmental, and industrial fuel combustion operations, on-site incineration and open burning, or transportation sources. Emission from area sources is calculated from stationary and mobile fuel combustion and from the waste-disposal data apportioned to an area. This classification relates to the multiple source classification of Table 2–2. Sources of information for this type of data are listed in Table 9–1.

A second method of tabulation of data is based on *point sources*, such as major fuel users (especially power plants) and industrial process sources, which are readily pinpointed on a map. These are defined as stationary sources that emit 100 or more tons per year of any single pollutant. One exception is aircraft emissions, which occur in the immediate vicinity of an airport and are sometimes considered as a point source. This classification relates to the specific source classification of Table 2–1.

stationary combustion sources

Stationary combustion sources generally use coal and residual and distillate fuel oil and gas for fuel. Wood may also be considered when burned in significant amounts. Emission rates are influenced not only by the quantity of fuel burned, but also by the chemical composition of the fuel, such as the ash or sulfur content.

Fuel consumed should be listed by *type of user* (manufacturing, steam electric utility, domestic, institutional, commercial), by *type of source* (i.e., area or point source), by *type of use* (i.e., process need or space heating), and finally by *geographical distribution* within the study area. When seasonal variations occur in fuel-use rates, they also affect the study and should be noted. Breakdown of fuel consumption by user category allows the use of emission factors developed specifically for each source category. *Total annual fuel consumption* data on individual fuels can be obtained from sources listed in Table 9–1. *Annual consumption of each fuel by user category* can also be obtained from listed sources for both manufacturing and steam electric utilities.

Annual consumption of each fuel by each point source may be obtained through mailed questionnaires based on mailing lists obtained from local and industrial trade directories. If questionnaires are not returned, an engineering estimate is made of a point source based on comparison with a similar source. The list may be refined by ground or aerial reconnaissance. The questionnaires request information on fuel combustion, solid-waste disposal, and process emissions. Fuels consumed are subdivided into process-need (fuel consumed for

TABLE 9-1   Reference Guide

Data Items	Sources of Information
Meteorological data (degree days)	U.S. Weather Bureau, "Local Climatological Data" (a)
Geographic location of industries	Aerial photographs, land-use maps, local building and fire depts., Directory of Manufacturers
Geographic distribution of domestic fuel use	Land-use maps, U.S. Bureau of Census "Census of Housing, HC(1)," "Number of Heating Units by Census Tracts," "PHC(1)–Census Tracts", local gas utilities company, local distributors of fuels
Geographic distribution of commercial fuel use	Land-use maps, U.S. Bureau of Census, "PC-C General Economic and Social Characteristics," local planning agency
Geographic distribution of institutional fuel use	Local board of education, land-use maps
Total annual gasoline sales	"Petroleum Facts and Figures"(c), U.S. Bureau of Census, Census of Business, "Retail Trade"
Geographic distribution of traffic	Traffic flow maps
Seasonal distribution of traffic	Local traffic control agency
Diesel fuel consumption	Local traffic control agency, National Trucking Association, local transit companies
Total area estimate of refuse production	Local sanitation agency, private haulers of refuse
Refuse disposal at collective sites	Local sanitation agency, incinerator and dump operators
Location of industries	Directory of Manufacturers

TABLE 9-1   (Cont'd)

Data Items	Sources of Information
Production data	Individual industries
Chemical composition of fuels	"Mineral Year Book—Fuels," "Burner Fuel Oils—Mineral Industry Surveys" (d)

(a) Supt. of Documents, Govt. Printing Office, Wash., D.C. 20402.

(b) National Coal Association, 1130 Seventeenth St., N.W., Wash., D.C.

(c) American Petroleum Institute, 1271 Ave. of the Americas, N.Y., N.Y. 10020.

(d) U.S. Dept. of Interior, Bureau of Mines, Wash., D.C. 20240.

processing the product being produced) and space-heating fractions based on information obtained from each individual source.

Annual consumption of each fuel by area sources may be computed as follows: annual consumption of each fuel in each user category is divided into two source subgroups, point and area sources. Therefore, the difference between annual consumption of fuel by a given user category and that consumed by the point sources under the respective user category is the annual fuel use by area sources in that category. For example, if annual usage of coal by manufacturing category is 1,000,000 tons and three large manufacturing concerns (point sources collectively) use 800,000 tons/year, the annual consumption by manufacturing area sources is 200,000 tons/year. Within the area source, process-need and space-heating fractions may be assumed to be subdivided in a manner similar to that determined from point sources. Presuming that the point source data yielded the information that of the 800,000 tons used by point sources 600,000 was used for process needs and 200,000 for space heating, the remaining 200,000 tons burned by area sources can then be broken down as follows:

$$200{,}000 \text{ tons} \times \frac{600{,}000}{800{,}000} = \begin{array}{l} 150{,}000 \text{ tons used for } process \\ needs \text{ by area sources} \end{array}$$

$$200{,}000 - 150{,}000 = \begin{array}{l} 50{,}000 \text{ tons used for } space\ heating \\ \text{by area sources} \end{array}$$

Commercial and institutional usage should be considered as used for space heating. Domestic usage of fuels is for space heating, water heating, and cooking. Some computations on household usage of fuels are covered later in this chapter.

Determination of daily fuel-use rate, corresponding to minimum, average, and maximum space-heating-demand day can now be determined.

The fuel for processes is assumed to be used at a uniform daily rate, with the daily rate being equal to 1/365 of the annual fuel consumption. The degree-day data required for calculating space-heating-demand rates are

1/ Number of days per year showing a degree-day value.

2/ Total number of degree days per year.

3/ The annual maximum degree-day value.

[A degree-day value represents 1° declination from 65°F in mean ambient air temperature for 1 day (data obtainable from U.S. Weather Bureau).]

Minimum space-heating-demand day: The minimum degree-day value is zero and occurs during the summer months. The space-heating fuel use is zero and only process fuel use is considered.

Average space-heating-demand day: Annual fuel consumption for space heating divided by the number of days showing a degree-day value yields the rate of fuel consumption for space heating corresponding to the average degree day.

Maximum space-heating-demand day: Space-heating fuel use on maximum demand day is determined as follows:

$$\frac{\text{max degree-day value}}{\text{total no. degree days/yr}} \times \text{annual space-heating fuel use}$$

Problem: Assume annual coal consumption by manufacturing area source:

150,000 tons for process needs

50,000 tons for space-heating needs

From meteorological data:

days showing a degree-day value =   250

total no. degree days/yr   = 6,000

maximum degree-day value   =   60

Fuel-consumption rates for the three conditions are

minimum day: $\dfrac{150,000}{365} + 0 = 410$ tons/day

average day: $410 + \dfrac{50,000}{250} = 410 + 200 = 610$ tons/day

maximum day: $410 + \dfrac{60}{6,000}(50,000) = 410 + 500 = 910$ tons/day

Similar calculations are performed for each fuel by user category and each point source. Results can be presented in tabular form.

mobile combustion sources

Fuels used by mobile combustion sources generally include gasoline and diesel for automobiles, buses, trucks, locomotives, airplanes, and ships. The latter three are normally excluded from the air pollution inventory but can be a major contributing factor in some areas. Gasoline consumption data for automobiles and trucks are usually based on total annual sales of gas in the area, traffic-flow maps, and an estimate of seasonal variations in traffic counts. Gasoline consumption may be allocated on a vehicle mile basis. Gasoline sales data may be obtained from the State Petroleum Marketers Association or other sources listed in Table 9–1. Assuming the ratio of gasoline sales to total service-station sales is comparatively constant within a state, you may estimate area gasoline sales as follows:

$$\frac{\text{service-station sales in study area (\$)}}{\text{service-station sales in state (\$)}} \times \text{gas sales in state (gal)}$$

$$= \text{gas sales in study area (gal)}$$

A check on those figures may be made by multiplying the number of vehicles registered times the average vehicle mileage per auto year and dividing by the average miles per gallon to obtain an estimate of yearly gas consumption.

Local traffic-control agencies usually have traffic-flow maps and traffic counts that may be used for computing estimates of gasoline usage by automobiles and trucks. Diesel fuel in trucks and buses is usually calculated on a basis of vehicle miles traveled in the area times 5.1 miles/gal. Distribution to reporting zones can be assumed to be the same as for gasoline.

refuse combustion sources

Solid-waste disposal should no longer be a major contributing factor since most communities have laws that prohibit open burning. If landfills where no burning is allowed and incinerators with proper control devices are used, pollution from these sources should be minimal. However, emission factors are available in Public Health Service Publication No. 999-AP-42 for assessing the importance of these sources.

industrial process loss sources

The quantities of the different contaminants discharged from most industrial and commercial establishments are attributable to two general types of operations: first, the *pollutants generated by the combustion of fuels* for space heating and process needs, and, second, the *pollutants generated by the industrial processes themselves*. These processes may contribute to localized air pollution problems in the vicinity of an individual plant without significantly polluting the community's total air supply. Emission rates of pollutants from process losses may be estimated by considering the types of processes and materials used, the production volume, and the efficiency of air pollution control equipment of each industrial process individually; the appropriate emission factors are then applied.

emission factors

The *emission factor* is a statistical average of the mass of pollutant emitted from each source of pollution per unit quantity of material handled, processed, or burned. In mobile sources, emission factors are usually based on miles of travel. Emission factors are determined by source sampling, by analyzing data provided by plant management personnel, or by analyzing data reported in technical publications. The conversion of fuel, solid-waste, and industrial-process data into air pollutant quantities is accomplished through the use of emission factors. Emission factors, related mostly to production data, have been developed for many of the industrial operations that may be important in air pollution surveys. One source is the *Inventory of Air Contaminant Emissions* published by the New York State Air Pollution Control Board.

The most comprehensive compilation of emission factors can be found in Public Health Service Publication No. 999-AP-42. This publication is available from the Superintendent of Documents, U.S. Government Printing Office, Washington, D.C., for 35 cents per copy. Although these emission factors represent the best available data at the time of publication, they are subject to continually changing conditions. Moreover, these factors should be used only if local data cannot be obtained from specific sources in the area of concern. Some emission factor data not included in PHS Pub. No. AP-42 are incorporated in this textbook, as well as a few tables extracted from AP-42 for use in computing a limited number of problems.

COMPUTATION OF POLLUTANT QUANTITIES

Having determined the pollutants of concern in the area and located the sources of these pollutants, the next step is to determine the average mass of pollutants emitted from each source (source sampling or use of emission factor tables). Finally, the quality and daily quantity of materials handled, processed, or burned by each source are determined and the rate (weight in units per day) at which each pollutant is emitted to the community atmosphere is tabulated.

In computations related to stationary combustion sources, some fuel properties that must be considered are listed in the following tables: Table 9-2 lists heating values and sulfur content of selected coals, Table 9-3 gives fuel-oil average properties, and Table 9-4 gives an analysis of the components of natural gas.

TABLE 9-2   Heating Values and Sulfur Content
of Selected Coals

Rank	State	BTU/lb	Ash %	Sulfur %	Carbon %
Anthracite					
Meta-anthracite	Rhode Island	9,620	28.3	- -	67
Anthracite	Pennsylvania	13,500	8.5	0.5	84
Bituminous					
Low-volatile	West Virginia	14,550	3.9	0.6	75
High-volatile A	Pennsylvania	13,610	9.1	1.3	58
High-volatile B	Kentucky	12,950	4.2	2.6	49
High-volatile C	Illinois	11,480	8.6	4.3	39
Lignitic					
Lignitic	North Dakota	7,210	6.2	1.7	31

TABLE 9-3   Fuel Oil—Average Properties

Grades	Gravity API(a)	Weight per volume Lbs./Gal.	Kilograms/Gal.	Sulfur Content(a) (Decimal)	BTU/ 10^6 Bbl(b)
1 (c)	42	6.79	3.08	.0013	5.8
2 (d)	35	7.08	3.21	.0035	5.8
4 (b)	21	7.73	3.50	.009	5.8
5 (e)	15	8.04	3.65	.013	6.3
6 (e)	10	8.33	3.78	.016	6.3

(a) Data obtained from Burner Fuel Oils, 1958, Petroleum Products, Survey No. 6, U.S. Dept. of Interior, Bureau of Mines, Oct. 1958.

(b) 1 barrel (bbl) = 42 gallons.

(c) Special use oils.

(d) Distillate oil used most for domestic and space heating.

(e) Residual oil used for commercial, mfg., and electric generators.

These factors will vary for different localities, based on fuels used, and the best local data should be used. As an example, No. 1 oil may have sulfur content of ½% and No. 6, 1.8 to 2.1% with BTU values from 137,000/gal. for No. 1 to 152,000/gal. for No. 6 oil.

calculating domestic fuel use

The Census of Housing records the number of dwelling units heated with various types of fuels for a given area. To convert this information into domestic fuel use, certain assumptions are made:

1/ Average energy use for space heating in this country is 70×10^6 Btu/yr-household.

2/ Average number of annual heating degree days for the country is 4,600. A heating degree day is a unit representing 1° of declination from a given point (65°F) in the mean ambient air temperature for 1 day. This standard unit is reported by the U.S. Weather Bureau. For example, if the average daily temperature is 50°F, the number of heating degrees for that day is 65 − 50 = 15.

3/ Average size of household: 4.9 rooms/dwelling unit. (Use 5 rooms/dwelling unit.)

4/ Assumed fossil-fuel characteristics (based on experience factors for fuels obtained from census data):

TABLE 9-4   Natural Gas—Analysis of Components

| Gas | Constituents of Gas, % by Volume | | | |
	CO	N_2	CH_4	C_2H_6
Natural gas, Texarkana	0.80	3.20	96.00	- -
Natural gas, Cleveland	- -	1.30	80.50	18.20
Natural gas, Oil City, Pa. (a)	- -	1.10	67.60	31.30

(a) Typical sulfur content of natural gas used in the Pennsylvania area is 0.034 to 0.035%. Natural gas, wet – 1,075 BTU/ft^3; natural gas, dry – 1,050 BTU/ft^3; manufactured gas – 550 BTU/ft^3

Fuel	Heating Value	Overall Efficiency of Heating Plant (%)
Coal	26×10^6 Btu/ton	50
Oil	145,000 Btu/gal	60
Gas	1,000 Btu/ft^3	75

(*Note*: More specific heating values may be computed from data in Table 9–2 if the type of coal used in the area is determined.)

5/ Heating requirements per household:

$$\frac{70 \times 10^6}{4,600} = 15,200 \text{ net Btu/household-degree heat}$$

6/ Fuel requirements per household:

coal: $\frac{15,200}{(0.50)\ 26 \times 10^6} = 0.0012$ ton coal/household-degree day

oil: $\frac{15,200}{(0.60)\ 145,000} = 0.18$ gal/household-degree day

gas: $\frac{15,200}{(0.75)\ 1,000} = 20.27$ ft^3/household-degree day

7/ Summary of estimating factors:

> coal: 0.0012 ton/household-degree day
>
> oil: 0.18 gal/household-degree day
>
> gas: 20.27 ft³/household-degree day
> (based on 5 rooms/dwelling unit)

8/ Problem: area data are as follows:

Fuel	No. Dwelling Units	Rounded
Coal	458,974	460,000
Oil	332,125	335,000
Gas	346,125	350,000

Average size of dwelling: 4.4 room/units.
Average number of annual heating degree days: 6,113.

calculating domestic coal use

$$460,000 \times 0.0012 \times 6,113 = 3,380,000 \text{ tons coal/yr}$$

Correction for number of rooms:

$$3,380,000 \times \frac{4.4}{5} = 2,970,000 \quad \text{or} \quad 3,000,000 \text{ tons coal/yr}$$

For those rare cases where wood is used as fuel, the following data may be useful. Wood (lb/ton of wood burned) emission factors are as follows:

Particulates	10
SO_2	Negligible
CO	30 to 65
NO_x	11

All wood, including mill wastes, produces 20,960,000 Btu/cord.

calculating industrial fuel use

Example of coal combustion by a power plant burning 100,000 tons coal/yr in a steam generator of 450×10^6 Btu/h capacity:
Find: Annual HC emissions (refer to Table 9–5).

TABLE 9-5   Gaseous Emission Factors for Coal Combustion
(Pounds per Ton of Coal Burned)

Pollutant	Type of Unit		
	Power Plant	Industrial	Domestic and Commercial
Aldehydes (HCHO)	0.005	0.005	0.005
Carbon monoxide	0.5	3.0	50.0
Hydrocarbons (CH$_4$)	0.2	1.0	10.0
Oxides of N (NO$_2$)	20.0	20.0	8.0
Oxides of sulfur (SO$_2$)	38S (a)	38S (a)	38S (a)

Extracted from AP-42, corrected by AP-66.

(a) S equals % sulfur in coal; e.g., if sulfur content is 2%, the oxides of sulfur emission would be 2 x 38 or 76 pounds of sulfur oxides per ton of coal burned. 5% of S remains in the ash.

$$(100,000 \ \text{tons/yr}) (0.2 \ \frac{\text{lb HC}}{\text{tons coal}}) = 20,000 \ \text{lb HC/yr}$$

Example of fuel oil combustion by a power plant burning 50,000,000 gal oil/yr in a steam generator of 250×10^6 Btu/yr capacity:

Find: Annual CO emissions (refer to Table 9–6).

TABLE 9-6   Emission Factors for Fuel-Oil Combustion[a]
(1,000 gal of oil burned)

Pollutant	Type of Unit			
	Power Plant Residual[b]	Industrial and Residual	Commercial[c] Distillate	Domestic Distillate[d]
Aldehydes (HCHO)	0.6	2	2	2
Carbon monoxide	0.04	2	2	2
Hydrocarbons (CH$_4$)	0.8	2	2	3
Oxides of nitrogen (NO$_2$)	104.0	72	72	12 to 72
Sulfur dioxide	157S[e]	157S	157S	157S
Sulfur trioxide	2.4S	2S	2S	2S
Particulates	10.0	23	15	8

a. Extracted from *Compilation of Air Pollutant Emission Factors,* PHS (NAPCA), Pub. No. 999-AP-42. Washington, D.C.: U.S. Government Printing Office, 1968. Corrected by *Air Pollution Control Technique Manuals* (Carbon Monoxide, Nitrogen Oxide and Hydrocarbons from Mobile Sources), PHS (NAPCA), Pub. No. AP-66. Washington, D.C.: U.S. Government Printing Office, 1970.

b. Power plant—greater than 100×10^6 Btu/h.

c. Industrial—10 to 100×10^6 Btu/h (except residual oil less than 100×10^6).

d. Domestic—less than 10×10^6 Btu/h.

e. S equals percentage of sulfur in oil.

$$(50{,}000{,}000 \ \text{gal/yr}) \ (0.04 \ \frac{\text{lb CO}}{1{,}000 \ \text{gal}}) \ = \ 2{,}000 \ \text{lb CO/yr}$$

Example of gas combustion by a power plant burning 50,000,000 ft³ natural gas/yr in a steam generator of 500×10^6 Btu/h capacity. *Find*: Annual NO_x emissions (refer to Table 9–7).

Table 9-7 Emission Factors for Natural-Gas Combustion[a]
($lb/1{,}000{,}000 \ ft^3$ of natural gas burned)

	Type of Unit		
Pollutant	Power Plant	Industrial Process Boilers	Domestic and Commercial Heating Units
Aldehydes (HCHO)	1	2	n
Carbon monoxide	n[b]	0.4	0.4
Hydrocarbons	n	n	n
Odides of nitrogen (NO_2)	390	214	116
Oxides of sulfur (SO_2)	0.4	0.4	0.4
Other organics	3.0	5.0	n
Particulates	15.0	18.0	19.0

a. Extracted from *Compilation of Air Pollutant Emission Factors*, PHS (NAPCA), Pub. No. 999-AP-42. Washington, D.C.: U.S. Government Printing Office, 1968. Assumes S content of 0.14/100 ft³. For process gas use $2.86C \ 16 \ SO_2/10^6 \ ft^3$, where C = grains of S/100 ft³ of gas.

b. n, negligible.

$$(50{,}000{,}000 \ ft^3) \ (390 \ \frac{\text{lb } NO_x}{1{,}000{,}000}) \ = \ 19{,}500 \ \text{lb } NO_x/\text{yr}$$

Example of coal combustion by a power plant burning 10,000 tons/yr in a spreader without fly-ash reinjection with 85 percent efficient multiple cyclone. Coal being used has 10 percent ash content. Emission factor equals $13A$ lb/ton of coal, where A equals ash content. *Find*: Annual particulate emissions (refer to Tables 9–8 and 9–9).

$$(10{,}000 \ \text{tons/yr}) \ (13 \times 10) \ (1 - 0.85) = 195{,}000 \ \text{lb}$$
$$\text{particulates/yr}$$

Example of coal combustion by an industrial source burning 10,000 tons/yr of West Virginia low-volatile bituminous coal. Emission factor is $38S$ lb/ton of coal, where S equals sulfur content: *Find*: Annual SO_2 emissions (refer to Tables 9–5 and 9–2).

$$(10{,}000) \ (38 \times 0.6) = 228{,}000 \ \text{lb } SO_2/\text{yr}$$

TABLE 9-8   Range of Collection Efficiencies for Common
Types of Fly Ash Control Equipment

| Type of Furnace | *Range of Collection Efficiencies, %* | | | |
	Electrostatic Precipitator	*High-Efficiency Cyclone*	*Low-Resistance Cyclone*	*Settling Chamber Expanded Chimney Bases*
Cyclone furnace	65-99 (a)	30-40	20-30	- -
Pulverized unit	80-99.9 (a)	65-75	40-60	- -
Spreader stoker	- -	85-90	70-80	20-30
Other stokers	- -	90-95	75-85	25-50

Extracted from AP-42.

(a) High values attained with high-efficiency cyclones in series with electrostatic precipitators.

regional estimate of emission factors for mobile sources

Assuming that the *gasoline-powered-vehicle* makeup in the area of concern is the same as the average national makeup (i.e., with respect to average vehicle age, make, engine displacement, etc.) and that the relative travel of vehicles is the same as the national average, the average national emission factors will apply equally well on a regional basis. All that is needed are vehicle miles of travel and percentage of travel that is urban.

The following equations can be applied to obtain regional emission estimates (from PHS Pub. No. AP-66):

$$TE = UE + RE$$

$$UE = (UF)\ (VMT)\ (a)\ (k)$$

$$RE = (RF)\ (VMT)\ (1 - a)\ (k)$$

TABLE 9-9 Particulate Emission Factors for Coal Combustion Without Control Equipment

Type of Unit	Particulate per Ton of Coal Burned, Lb.(a)	Percent 44 Microns or Greater	Percent 20 to 44 Microns	Percent 10 to 20 Microns	Percent 5 to 10 Microns	Percent less than 5 Microns
Pulverized						
General	16 A	25	23	20	17	15
Dry bottom	17 A	25	23	20	17	15
Wet bottom without fly ash reinjection	13 A	25	23	20	17	15
Wet bottom with fly ash reinjection (b)	24 A	25	23	20	17	15
Cyclone	2 A	10	7	8	10	65
Spreader stoker						
Without fly ash reinjection	13 A	61	18	11	6	4
With fly ash reinjection (b)	20 A	61	18	11	6	4
All other stokers	5 A	70	16	8	4	2
Hand-fired equipment	20	- -	- -	- -	- -	100

Extracted from AP-42.

(a) The letter A on all units other than hand-fired equipment indicates that the percent ash in the coal should be multiplied by the value given. Example: If the factor is 17 and the ash content is 10 percent, the particulate emission before the control equipment would be 10 times 17, or 170 pounds of particulate per ton of coal.

(b) Values should not be used as emission factors. Values represent the loading reaching the control equipment always used on this type of furnace.

where

TE = total emissions, tons/yr

UE = urban emissions, tons/yr

RE = rural emissions, tons/yr

UF = urban emission factor, g/mile

RF = rural emission factor, g/mile

VMT = vehicle miles of travel, miles/yr

a = fraction of travel that is urban

k = 1.1023×10^{-6} ton/g (a conversion factor)

solving the problem for the year 1975

Assume: 70,000 gasoline-burning automobiles were registered in area in 1968. No future travel estimates nor base-year travel figures are available; therefore, factor 9,400 (miles/yr/vehicle registered) will be used to compute VMT.

$$70,000 \times 9,400 = 658 \times 10^6 \quad VMT \text{ for } 1968$$

$$658 \times 10^6 \times 1.26 \text{ (from Table 9–10)} =$$
$$829 \times 10^6 \quad VMT \text{ for } 1975$$

TABLE 9-10 Factors for Estimating Vehicle Miles of Travel

Year of Interest	Base Year		
	1960	1965	1968
1960	1.00	0.82	0.75
1965	1.23	1.00	0.91
1968	1.34	1.10	1.00
1970	1.43	1.17	1.06
1975	1.69	1.38	1.26
1980	2.00	1.63	1.49
1985	2.37	1.93	1.77
1990	2.82	2.30	2.11

It is estimated that 0.80 is urban travel. Enter Table 9–11 and extract:

For urban travel (g/vehicle mile)	For rural travel (g/vehicle mile)
HC 7.39	4.78
CO 40.9	17.5
NO_x 6.84	7.81

Compute for HC:

$$UE = (7.39)\ (829 \times 10^6)\ (0.80)\ (1.1023 \times 10^{-6}) =$$
$$5.391 \times 10^3 \text{ tons/yr}$$

$$RE = (4.78)\ (829 \times 10^6)\ (0.20)\ (1.1023 \times 10^{-6}) =$$
$$0.894 \times 10^3 \text{ tons/yr}$$

$$TE = 5.39 \times 10^3 + 0.894 \times 10^3 =$$
$$6.285 \times 10^3 \text{ tons/yr HC}$$

(Divided by 365 days per year equals 17.2 tons/day **HC**)

TABLE 9-11   Average On-the-Road Emission Rates for Gasoline-Powered Motor Vehicles(a) (Grams/Vehicle-Mile)

Year	Hydrocarbon		CO		NO_x	
	Urban	Rural	Urban	Rural	Urban	Rural
1960	21.8	14.6	87.1	34.9	5.72	6.39
1965	20.4	13.2	87.6	35.6	5.76	6.53
1968	17.9	11.4	80.9	33.0	5.92	6.70
1970	14.6	9.49	67.9	28.1	6.27	7.10
1975	7.39	4.78	40.9	17.5	6.84	7.81
1980	4.20	2.65	27.6	11.9	7.08	8.18
1985	3.71	2.18	25.7	10.7	7.08	8.27
1990	3.69	2.18	25.6	10.8	7.06	8.31

(a) Includes cars and trucks, and applicable control systems in use as of 1971.

Another approach in computing HC emissions is to use emission data from Table 9–12 that indicate an emission factor of 0.363 lb/vehicle/day from auto exhausts and 0.2 lb/vehicle/day from engine crankcase blow-by

```
    0.363
+   0.2
    0.563  lb/veh/day

           70,000  vehicles
2000 / 39410.000 / 19.7 ton/day    (However, these tables are
       2000                         based on automobiles
       19410                        without controls.)
       18000
        14100
        14000
         1000
```

Emissions from automobiles are highly variable, depending upon geographical location and local driving patterns. In high-altitude cities, emissions of HC are 30 percent greater, CO 60 percent greater, and NO_x 50 percent less than in low-altitude cities.

Model 1966 vehicles with exhaust control devices reduce HC 35 percent and CO 67 percent and increase NO_x 26 percent. Model 1971 vehicles have reduced output as follows from uncontrolled levels: HC 80 percent, CO 69 percent, and, in California only, NO_x 33 percent.

TABLE 9-12　　Emission Factors for Automobile Exhaust

Type of Emission	Emissions		
	Pounds per 1000 Vehicle-Miles	Pounds per 1000 Gallons of Gas	Pounds per Vehicle-Day
Aldehydes (HCHO)	0.3	4	0.007
Carbon monoxide	165.0	2300	4.160
Hydrocarbons (C)	12.5	200	0.363
Oxides of nitrogen (NO_2)	8.5	113	0.202
Oxides of sulfur (SO_2)	0.6	9	0.016
Organic acids (acetic)	0.3	4	0.007
Particulates	0.8	12	0.022

Extracted from AP-42, Uncontrolled Automobile Exhausts. Based on average route speed of 25 mph in urban areas. Estimated 3.25 trips per day of 8 miles in length each. Average automobile travels about 14.4 miles per gallon of gas consumed. Hydrocarbons add 0.2 pounds per vehicle-day for engine crankcase blowby. Assume S content of 0.07%.

Diesel engine vehicle emission factors are shown in Table 9–13. Note that emissions of CO and HC are lower than from gasoline engines; however, all others are correspondingly greater. It is estimated that 62.5 percent of particulates from diesel exhaust are less than 5

TABLE 9-13   Emission Factors for Diesel Engines
(Pounds per 1,000 Gallons of Diesel Fuel)

Type of Emission	*Emission Factor*
Aldehydes (HCHO)	10
Carbon monoxide	60
Hydrocarbons (C)	136
Oxides of nitrogen (NO_2)	222
Oxides of sulfur	40
Organic acids (acetic)	31
Particulates	110

Extracted from AP-42. Assume S content of 0.3%.

microns in size and that 37.5 percent are 5 to 20 microns in size. No
control systems have been developed for diesel exhaust emissions. Another way of computing diesel-fuel consumption is to multiply the
vehicle mileage by 5.1 miles/gal.

   Aircraft emission factors are shown in Table 9–14. The factors
shown are combined and averaged figures for emissions during all
phases of aircraft operations (i.e., taxiing, taking off, climbing out,
approaching and landing) that take place below the arbitrarily chosen

TABLE 9-14   Emission Factors for Aircraft Below 3500 Feet
(Pounds per Flight) (a)

Types of Emission	*Jet Aircraft, Four Engine(b,c)*		*Turboprop Aircraft Engines*		*Piston-Engine Aircraft Engines*	
	Conventional	*Fan Jet*	*2*	*4*	*2*	*4*
Aldehydes (HCHO)	4	2.2	0.3	1.1	0.2	0.5
Carbon monoxide	35	20.6	2.0	9.0	268.0	652.0
Hydrocarbons (C)	10	19.0	0.3	1.2	50.0	120.0
Oxides of N (NO_2)	23	9.2	1.1	5.0	12.6	30.8
Particulates	34	7.4	0.6	2.5	0.6	1.4

Extracted from AP-42 corrected by AP-66.

(a) A flight is defined as a combination of a landing and a take-off.

(b) No water injection on take-off.

(c) For 3 engine aircraft multiply by 0.75, for 2 engine aircraft by 0.5, and for 1 engine
   aircraft by 0.25.

altitude of 3,500 ft. Emissions at cruise altitude above 3,500 ft are not of concern in conducting an emission inventory.

In some areas with heavy train travel, air pollution emissions from trains should be calculated. Water vessels are a major factor, too, if the area has considerable water travel within its boundaries.

PRESENTATION OF RESULTS

Following data gathering and emission calculation, the results are tabulated according to source and pollutant calculation. Table 9–15 is an example. Data for computation of other sources shown in Table 9–15 are included in PHS Pub. No. AP-42, which contains emission factors on fuel combustion, transportation, refuse incineration, chemical-process industries, food and agricultural industries, metallurgical industries, mineral-products industries, petroleum refineries, pulp and paper industries, and solvent evaporation and gasoline marketing. Another set of tables covering emissions from industrial operations and processes is available in *Inventory of Air Contaminant Emissions* (New York State Air Pollution Control Board), classified by operations and/or processes or classified by industries.

In addition to the tables summarizing the emissions of each pollutant by source category, data may be tabulated as emission density maps for each pollutant, as well as average day, average summer day, and average winter day emissions for each grid or each jurisdictional area. These may be prepared as overlays for imposition on the area map.

INTERPRETATION OF DATA

Maximum emissions result from normal activities with no controls. Minimum emissions result from application of the most advanced control techniques; however, this is not a static situation. Since a source–emission inventory provides no indication of severity or seriousness of effect upon the community for any of the pollutants inventoried, the relative magnitude of emission among the various pollutant classifications has no meaning. However, the inventory does provide an indication of the relative source contribution of a given pollutant to the community atmosphere; therefore, it does provide guidance and direction in the design of a control program.

TABLE 9-15   Emission of Air Contaminants to the Atmosphere by Source, in Tons per Day

Source	Organic Gases			Aerosols	Inorganic Gases			
	Hydro-carbons	Aldehydes & Ketones	Other Organics		Oxides of Nitrogen	Oxides of Sulfur	Carbon Monoxide	Other Inorganics
Transportation								
Automobiles (gasoline)								
Exhaust	680	10	12	25	250	19	6850	n
Blowby	140	u	u	u	u	n	n	n
Evaporation	145	0	0	0	0	0	0	0
Trucks & Buses (gasoline)								
Exhaust	130	3	4	8	80	6	2100	n
Blowby	40	u	u	u	u	n	n	n
Evaporation	45	0	0	⁻0	0	0	0	0
Trucks & Buses (Diesel)	8	n	n	2	8	2	2	n
Ships and Railroads	n	n	n	n	n	n	n	n
Aircraft								
Jet	n	n	n	1	1	0	1	0
Piston	15	n	n	n	5	n	110	n
Petroleum								
Refining	90	2	n	5	1	40	700	4
Marketing	100	0	0	0	0	0	0	0
Production	60	n	0	0	u	n	u	0
Organic Solvent Users								
Surface Coating	190	25	35	7	0	0	0	0
Degreasing	20	n	40	0	0	0	0	0

	1	2	3	4	5	6	7	8
Dry Cleaning	20	n	8	n	0	0	0	0
Plastics and Rubber	17	4	3	n	0	0	0	0
Other	50	14	13	n	0	0	0	0
Combustion of Fuel								
Coal	3	1	0	80	20	200	10	0
Liquid	2	1	6	38	187	454	n	0
Gaseous	4	1	2	14	150	13	1	0
Chemical								
Sulfur Plants	0	0	0	0	n	30	0	0
Sulfuric Acid Plants	0	0	0	n	n	20	0	0
Other	16	8	16	8	n	1	0	1
Metals	0	0	0	6	3	n	180	0
Incineration								
Municipal	n	n	3	3	3	1	3	n
Industrial	n	0	1	n	1	n	1	n
Other	0	0	n	1	n	n	n	0
Minerals	2	0	0	4	6	0	0	0
Miscellaneous	u	0	u	2	0	u	0	u
TOTALS—Rounded	1775	70	145	195	525	775	9960	5

n – negligible (0.05-0.5 ton per day)
u – unknown

REFERENCES

U.S. Department of Health, Education, and Welfare, *Air Pollution Control Technique Manuals*, PHS (NAPCA), Pub. Nos.

> AP-51, Particulates, 1969.
> AP-52, Sulfur Oxide, 1969.
> AP-65, Carbon Monoxide from Stationary Sources, 1970.
> AP-66, Carbon Monoxide, Nitrogen Oxide and Hydrocarbons from Mobile Sources, 1970.
> AP-67, Nitrogen Oxide from Stationary Sources, 1970.
> AP-68, Hydrocarbons and Organic Solvent from Stationary Sources, 1970.

Washington, D.C.: U.S. Government Printing Office.

U.S. Department of Health, Education, and Welfare, *Atmospheric Emissions from Coal Combustion—An Inventory Guide*, PHS (NAPCA), Pub. No. 999-AP-24. Washington, D.C.: U.S. Government Printing Office, 1966.

U.S. Department of Health, Education, and Welfare, *Calculating Future Carbon Monoxide Emissions and Concentrations from Urban Traffic Data*, PHS (NAPCA), Pub. No. 999-AP-41. Washington, D.C.: U.S. Government Printing Office, 1967.

U.S. Department of Health, Education, and Welfare, *Compilation of Air Pollutant Emission Factors*, PHS (NAPCA), Pub. No. 999-AP-42. Washington, D.C.: U.S. Government Printing Office, 1968.

U.S. Department of Health, Education, and Welfare, *A Rapid Survey Technique for Estimating Community Air Pollution Emissions*, PHS (NAPCA), Pub. No. 999-AP-29. Washington, D.C.: U.S. Government Printing Office, 1966.

RECOMMENDED FILMS

MA-45   Calculation of Estimated Emissions (20 min)
MA-40   Source Inventory
    Available: Distribution Branch
                National Audio-Visual Center (GSA)
                Washington, D.C. 20409

QUESTIONS

1/ What is included in a source–emission inventory?
2/ What is the basis used for selecting reporting zones?
3/ What is the difference between an area source and a point source?

4/ What is the usual source of information for calculating point source emissions?

5/ What are the three categories of data used to calculate area source emissions?

6/ What are emission factors and what publication contains the most comprehensive computation of emission factors?

7/ What type coal has the lowest sulfur content?

8/ Find annual CO emissions for a power plant burning 100,000 tons of coal/yr in a steam generator of 450×10^6 Btu/h capacity.

9/ Find annual SO_x ($SO_2 + SO_3$) emissions for a power plant burning 50,000,000 gal of fuel oil/yr in a steam generator of 250×10^6 Btu/h capacity.

10/ Find annual particulate emissions for 10,000 diesel trucks in the area of concern with distribution of diesel fuel reported as 10,000,000 gal/yr.

11/ Find annual HC emissions in the vicinity of a major airport recording an average of 500 flights/day. Of these 500 flights, 300 are four-engine conventional jet, 100 two-engine fan jet, and 100 two-engine piston aircraft.

12/ What is the difference between process need and process loss?

10

AIR POLLUTION CONTROL

OF

FIXED SOURCES

Student Objectives

—*To develop an understanding of control techniques used to abate air pollution from fixed sources.*
—*To learn what devices may be applied to control gaseous pollutants.*
—*To learn what devices may be applied to control particulant pollutants.*

APPROACHES TO AIR POLLUTION CONTROL

Air pollution control of fixed sources may be accomplished by two fundamental approaches, which are categorized as control by dilution in the atmosphere by dispersion, or controls at the source designed to reduce the air pollution emitted to the bare minimum.

control by dilution in the atmosphere by dispersion

The most positive way to abate air pollution is to prevent pollution from coming into existence. However, smokestacks are used to provide for a reduction of ground-level concentration of pollutants by giving natural atmospheric turbulence an opportunity to dilute the pollutant before it reaches ground-level receptors in harmful concentrations. This approach controls emissions to some degree and may help to obtain desired air quality, since the atmosphere has tremendous powers to dilute, disperse, and destroy a large variety of substances that man elects to discharge into it.

The effective approximate stack height is a minimum two and one-half times the height of the tallest building in the vicinity. This height should allow dispersion of the plume downwind five to ten times the height of the building. This effective height also depends upon wind factors, dilution, diffusion, and prevailing temperature (which are discussed in Chapter 12). A rough estimate of stack cost is 200×10^3 dollars for a 1,000-ft stack.

Community planning should include air zoning to prevent harmful ground concentrations from occurring within designated areas. Plans should require the location of plants where stacks will be most efficient and where fewer people will be affected by pollution from the stacks. Meteorological studies of a community can assist in air zoning. Data on approaching or expected adverse weather conditions that might require stack emissions to be halted to prevent an air pollution episode can be obtained from daily meteorological reports.

Smokestacks are unsightly and sometimes provide a hazard to low-flying aircraft. Stacks are no longer a symbol of industrial progress nor are they very efficient in reducing air pollution unless accompanied by one or more of the source controls covered in the following paragraphs.

control at source

Control at source may be accomplished by keeping the pollutant from coming into existence or by destroying, altering, or trapping it before it reaches the atmosphere.

Source relocation is one method of control at source. In the study of meteorological effects in community air zoning it is sometimes possible to determine a more satisfactory location for an industry that is causing unacceptable air pollution in its present location. By relocating away from heavily populated areas and taking advantage of prevailing winds, an acceptable level of air pollution may be reached.

Source shutdown is another method of control at source. A source may be shut down for a period of time when air pollution levels threaten public health. The Federal Environmental Protection Agency as well as some state air pollution control agencies have requested source shutdowns when expected adverse weather conditions threatened an air pollution episode. When a source failed to comply with such a request, federal or state agencies have obtained court orders to force shutdown when public health was threatened.

Fuel or energy substitution is another method of control at source. Fuel or energy substitution may be accomplished by replacing soft coal with hard coal, residual oil, distillate oil, or natural gas. An even more drastic improvement can be made by replacing fossil fuels with hydraulic, electric, nuclear, solar, or geothermal energy. Fuel can be treated before combustion by desulfurizing coal and fuel oil or by refining coal or natural gas to liquefied natural gas (LNG) or to liquid petroleum gas (LPG), which has sulfur removed.

Process changes may be put into effect that will conserve energy as well as reduce pollution as another method of control at source. For example, open-hearth furnaces in the steel industry have been replaced with controlled basic oxygen furnaces or electric furnaces that emit less pollution into the atmosphere by reduction of smoke, carbon monoxide, and metal fumes. Such changes, coupled with various combinations of gas-cleaning devices, can be very effective in reducing air pollution.

Good operating practices are another method of control at source. Irrespective of the type of equipment that is installed, the fuel burned, or the raw material used, the operator is the key to the ultimate reduction of air pollution from any given source. Equipment must be properly applied, installed, operated, and maintained to minimize the emission of pollutants. As an example, the introduction of excess liquid sulfur in the burner at a sulfuric acid plant, without enough excess air, may result in undue emissions of sulfur oxides.
Another example would be the excessive emissions of fly ash from a power plant due to the oversight of an operator by introducing too much excess air into the boiler furnace.
Failure to properly lubricate an exhaust fan on an apartment

incinerator could cause excessive air pollution due to a lack of combustion air and a safety problem due to smoke and fire going back up the charging chute.

The fuel industry, equipment manufacturers, and government agencies have prepared guidelines for good operating practices. The National Coal Association conducts courses for boiler operators using coal. Some cities, such as New York City, require furnace stokers to attend courses of instruction to certify that they are qualified in the best procedures.

Air pollution control devices or techniques may be applied as another method of control at source. The installation and operation of air pollution control devices or techniques are designed to destroy, mask, counteract, or collect pollutants. These devices or techniques are often needed in addition to other approaches to source control, covered in previous paragraphs, in order to attain the level of air quality desired.

Control devices are generally designed to control either gaseous pollutants or particulate pollutants, since few devices are efficient in the control of both. Where it is possible to do so, a device may be used to control both gaseous and particulate pollutants. Some devices are designed for a specific type of pollutant or a specific size of particulate. Some considerations involved in the selection of a particular device are: Is product recovery involved? Are both dusts and gases involved? What disposal facilities are involved? Is heat recovery a factor? Are there combining factors (e.g., in the case of a captive dust chemical by-product, is the recoverable by-product corrosive and does it require special handling)?

DEVICES AND TECHNIQUES FOR CONTROL OF GASEOUS POLLUTANTS

The technique and the device used to control a gaseous pollutant depend on the properties of the specific gas to be controlled. The technique or control methods are generally classified under one of the five different methods of treatment: absorption, adsorption, combustion, closed collection and recovery systems, and masking and counteraction. Although devices applying these methods are designed primarily to control gaseous emissions, some may also reduce visible emissions and particulates.

gas absorption

The principle of gas absorption is a gas–liquid contacting process for gas separation that utilizes the preferential solubility or chemical reactivity of the pollutant gas in the liquid phase. In the gas absorption technique, effluent gases (gases being emitted from the source) are passed through *absorbers* (scrubbers) containing *liquid absorbents* that remove, treat, or modify one or more of the offending constituents in the gas stream.

Efficiency depends upon the amount of surface contact between the gas and the liquid (since the greater the surface, the greater will be the absorption), the time gas is allowed to remain in contact with the liquid, the concentration of the absorbing media, and the speed of reaction between the absorbent and the gas.

Liquid absorbents may be classified as *reactive* if the absorbent utilizes chemical change to remove pollutants. As an example, sulfur dioxide may be removed from flue gases by injecting water and limestone, which reacts to form calcium hydroxide. Calcium hydroxide then reacts with sulfur dioxide to form calcium sulfate salt, which can be scrubbed from the gas stream by more water.

If gases are removed by simply dissolving the gas without chemical change, the absorbent is termed a *nonreactive* absorbent. Water or heavy carbon oil are examples of nonreactive absorbents.

An absorbent that cannot be regenerated for reuse but instead must be discarded is referred to as a *nonregenerative* absorbent. Water is one example. An absorbent that can be forced to release the gaseous pollutant that it has captured reversibly by application of heat or steam or by pressure change is referred to as a *regenerative* absorbent. A *regenerative* absorbent may allow reuse of expensive chemicals or catalysts, may be necessary to chemically neutralize the pollutant for disposal as a solid or liquid, or may aid in concentration of pollutant gas for further processing. An example of a regenerative absorbent is carbon tetrachloride, which under pressure combines with chlorine gas and removes it from the effluent gas stream. Then by varying temperature and pressure in a stripping tower the two can be separated again and the carbon tetrachloride, free of chlorine, can be used again as an absorbent. The chlorine in turn is recovered in a gaseous or liquid state for possible commercial usage.

Absorbers (scrubbers) are the devices that physically contain the absorbent liquid through which the effluent gas must pass. The ar-

rangement and design is devised to obtain maximum removal of pollutant gases from the effluent gas stream. Some of these absorber
types are

1/ A *packed tower* (Fig. 10–1) consists of a vertical shell filled with
a suitable packing material. One example of packing material is
polyethylene (Tellerette) which has the shape of a helix, that is,

Packed tower absorber

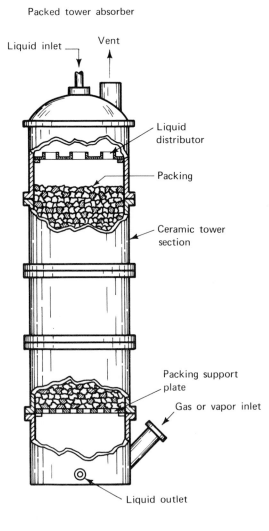

Fig. 10–1 Packed tower absorber (Courtesy of Mfg. Chem. Assoc. Inc.)

formed into a doughnut shape. These and other plastic packings are virtually unbreakable and their light weight permits design of smaller scrubbers. The absorbent flows over the surface of the packing material in thin films, thereby presenting a large liquid surface in contact with the gas. Usually the flow through a packed tower is countercurrent, with absorbent introduced at the top to trickle downward while gas is introduced at the bottom to pass upward through the packing. The packed tower is effective for removing mist (liquid particles 10 microns or smaller formed by condensation of molecules from the vapor state). For absorbing corrosive gases and vapors a packed tower is usually more economical to construct since less corrosion resistant materials are required. Packed towers operate under a relatively low pressure drop for equal operating conditions, and thus are more suitable for vacuum operation. Foamy liquids pass through packed towers better than through the plate towers described next.

2/ A *plate tower* (Fig. 10–2) consists of a vertical shell in which a large number of equally spaced circular perforated (sieve) plates are mounted. Gases and vapors bubble upward through the liquid seal above each plate. This passage of gases through the perforation prevents liquid from passing through the holes. At the side of each plate a conduit called a down spout is provided to pass the liquid downward from plate to plate. The plate tower, because of its design, is preferable to the packed tower

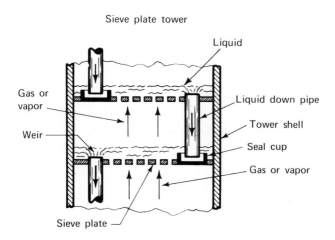

Fig. 10–2 Sieve plate tower (Courtesy of Mfg. Chem. Assoc. Inc.)

when liquid contains suspended solids or relatively insoluble offensive gases because it is easier to clean and can handle higher liquid flow rates. When heat of solution must be removed, the plate tower is preferable due to ease of installation of cooling coils. However, initial cost is higher for the plate tower than for the packed tower.

3/ A *spray tower* (Fig. 10–3) consists of a vertical chamber that utilizes the principle of interception—contact between the mist particle and spray droplet. The absorbent is sprayed through the effluent gas stream, which provides turbulence in the gaseous phase (around the outside of the liquid droplet), which is especially desirable when treating highly soluble offensive gases. By applying centrifugal force and liquid spray to the gas path at the same time, maximum contact between gas and liquid is possible. The spray tower is effective for removal of large liquid

SPRAY TOWERS

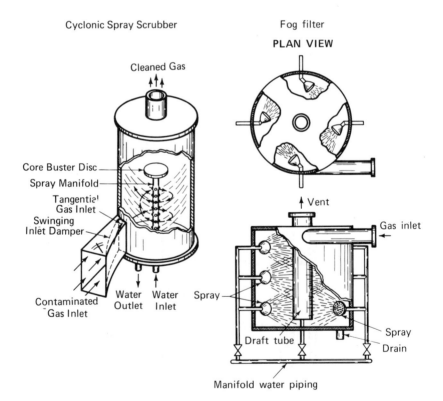

Fig. 10–3 Spray tower (Courtesy of Mfg. Chem. Assoc. Inc.)

particles greater than 10 μ in size and may also be used for particulate removal.

4/ A *liquid jet scrubber* (Fig. 10–4) is a two-chambered device in which the absorbent enters one chamber through the top as a jet spray against the effluent gas flow that is induced into the upper side. This jet spray atomizes the absorbent or generates minute liquid spray droplets that intercept smaller particles, thereby increasing collection efficiency. Noncondensable gases are expelled from the other chamber.

Water Jet Scrubber

Fig. 10–4 Water jet scrubber (Courtesy of Mfg. Chem. Assoc. Inc.)

5/ *Agitated tanks* (Fig. 10–5) contain a stirring device that causes turbulence by throwing absorbent and effluent gas against baffle plates on the side. This turbulence provides a more satisfactory absorption of gases by the liquid absorbent when effluent gases contain particulates as well as gaseous pollutants.

Some of the operating problems in the use of *all* absorbers are that temperature must remain below 100°C to keep the absorbent in a liquid state, which means stack gases cannot be treated without cooling. This destroys natural draft and imposes a large cooling load on the plant. Entrainment separators must be installed after most absorbers to prevent carryover of the absorbent and creation of a new pollution problem.

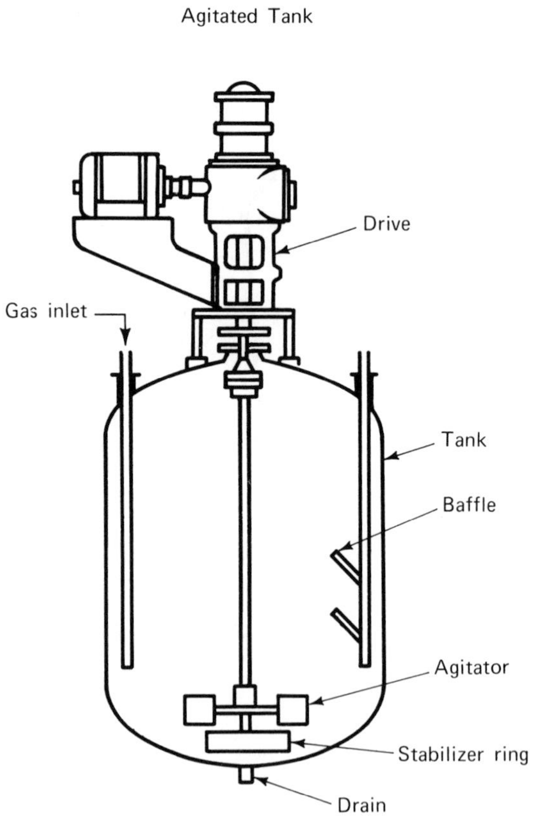

Agitated Tank

Fig. 10–5 Agitated tank (Courtesy of OMD-APCO-EPA).

Applications of absorber equipment include the removal of sulfur oxides in flue gases from coal-burning steam-boiler plants, sulfur oxides from smelter gases, sulfuric acid fumes from paint pigment manufacture, hydrogen sulfide from natural gas and petroleum refinery gases, chlorine gas from chemical processings, halogens, carbon dioxide, and particulates from laboratory hood effluent gas, nitrogen dioxide from factories producing nitric and sulfuric acid, and hydrogen chloride gas from plating processes.

gas adsorption

Gas absorption, previously discussed, is based on gases reacting with a *liquid absorbent*, which takes place when molecules or atoms are distributed in the bulk of the interacting phase. *Gas adsorption*, on the other hand, is based on gases reacting with a *solid adsorbent*, which takes place when the molecules or atoms sorbed are concentrated only at the interface of the solid. Gas adsorption is based on the principle of passing effluent gas through solid adsorbers contained in an adsorption collecting device. Adsorption may be physical or chemical in nature.

The efficiency of adsorption devices depends upon several factors. Physical adsorption will occur under suitable temperature and pressure conditions in any gas–solid system, whereas chemical adsorption takes place only if the gas is capable of forming a chemical bond with the surface. Furthermore, a physically adsorbed molecule can be removed unchanged at a reduced pressure at the same temperature at which the adsorption took place. The removal of the chemically adsorbed layer is far more difficult. The boundary layer between the gas to be adsorbed and the solid adsorbent is most important in the phase interaction. Therefore, it is expedient to create the maximum obtainable surface area within the solid phase. Also, some adsorbents are selective as to what gases they will preferentially adsorb.

Another factor to consider is the separation (desorption) of the gaseous pollutant, which has been adsorbed, from the solid adsorbent in order to regenerate the adsorbent for reuse as well as to dispose of the gas. Under some conditions it is not economically feasible to separate the adsorbent and the gas, but rather one must dispose of the entities in the combined state. In some instances the recovery of the gaseous pollutant itself is economically worthwhile.

Solid adsorbents of the industrial type are generally capable of adsorbing both organic and inorganic gases. However, their preferential adsorption characteristics and other physical properties make each one more or less specific for a particular application. As an example, *activated alumina, silica gel,* and *molecular sieves* (synthetic, silicate, or zeolite) will adsorb water preferentially from a gas phase mixture of water vapor and an organic contaminant. Alumina and silica gel are used industrially to dry gases. Bauxite is used in the treatment of petroleum fractions and drying of gases. *Activated carbon* preferentially adsorbs nonpolar organic compounds, since water molecules, being highly polar, exhibit strong attraction for each other, which competes with their attractions for the nonpolar carbon surface; as a result, the larger, less polar organic molecules are selectively adsorbed by the carbon. Activated carbon is most often used for removal of organic solvent vapors. (*Activated* carbon or charcoal is prepared by the use of steam at high temperatures, so that part of the carbon material is burned away to create a large internal pore structure and provide an exceptionally large internal surface area.) Activated carbon is the most-used adsorbent, and only devices designed for its use are discussed.

Adsorbers (Fig. 10–6) are the devices that physically contain the adsorbent solid through which the effluent gas must pass. Some of these adsorber types are

1/ *Thin-bed adsorbers* using activated charcoal as adsorbent in thin layers (½ in. thick) will save on power because of less resistance to the flow of air. The adsorbent may be contained in a canister design or in a folded-cell design. In purification of air being brought into a building from outside, the thin-bed adsorbent is most always used. The air, although at times highly polluted, at the same time is highly diluted, and pollutants are usually in trace quantities only. Adsorption is rapid, and contaminants cannot build up on the surface rapidly enough to reduce the collection efficiency of the thin-bed adsorber layer.

2/ *Deep-bed adsorbers* using activated charcoal as adsorbent in layers deeper than ½ in. will occupy the least amount of space and be simpler to fabricate than thin-bed adsorbers, given a definite amount of activated charcoal. Deep-bed adsorbers will be used where those savings on power costs are overriden by other determining factors. For instance, in purification of air being exhausted from a building, heavy concentrations of pollutants are involved. The buildup of contaminants on the adsorber might

Activated Carbon Adsorber

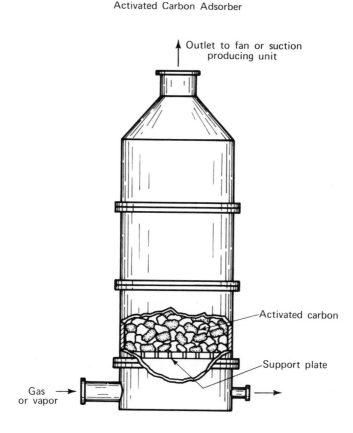

Fig. 10–6 Activated carbon adsorber (Courtesy of OMD-APCO-EPA).

be too rapid for a thin-bed adsorber to handle without becoming clogged.

3/ *Other designs* may include fixed-bed adsorbers in vertical cylinders or in horizontal cylinders, movable-bed adsorbers contained in a drum rotating within an enclosure, and adsorbers in series. Each design has some advantage in specific application under a given set of circumstances.

   Applications of adsorber equipment include recovery of isopropyl alcohol from a citrus-fruit processing plant, recovery of methyl chloroform from a movie-film processing plant, recovery of ethyl alcohol vapors from a whiskey warehouse, cleansing of air exhausted from kitchen exhaust hoods, and removal of contaminants from air prior to use in an operating room or an electronics control room.

combustion

Many organic compounds released from manufacturing operations can be converted to innocuous carbon dioxide and water by combustion (rapid oxidation). To obtain complete combustion, the proper proportion of oxygen, temperature, turbulence, and time (the three "T's" of combustion) must be provided.

Oxygen is necessary for combustion to occur. The end products of combustion depend on the supply of oxygen. For instance, when methane is burned with insufficient oxygen solid carbon results, forming particles of soot and smoke. With sufficient oxygen, the carbon is burned to carbon dioxide.

Temperature must be maintained at ignition temperature (temperature at which more heat is generated by the reaction than is lost to the surroundings). The ignition temperatures for combustion of combustible substances cover a large range: for example, sulfur, 470°F; charcoal, 650°F; methane, 1,170 to 1,380°F; carbon monoxide, 1,130 to 1,215°F. Insulation may be required to save loss of heat from the system. Proper stack height will help maintain stack gas temperatures higher than ambient air temperature, a condition that helps maximize dilution in the atmosphere.

Turbulence provides intimate mixing of the oxygen with the combustion substance at all times. Baffles or injection nozzles may be required to maintain the necessary turbulence. The shape and height of the stack may also be factors in providing needed turbulence.

Time for efficient burning must be provided by a combustion chamber of appropriate size. Increasing the height of the stack provides more time for burning and thereby reduces smoke emitted.

The type of combustion process to be used depends upon the type of fuel to be burned. However, whether fuels are solid, liquid, or gas, most contain C, H, O, and S, which produce CO_2, CO, SO_2, H_2O, and unburned HC. Energy produced by combustion may be measured in British thermal units (Btu's). One Btu represents the energy produced by one wooden match, the temperature required to raise 1 pint of water 1°F, or 778 ft-lb of work.

If a proper mix of oxygen, time, turbulence, and temperature are maintained, the combustion of 1 lb of carbon will produce 14,600 Btu. Combustion of 1 lb of hydrogen will produce water vapor with an equivalent 62,000 Btu. Combustion of 1 lb of sulfur produces sulfur dioxide with an equivalent of 4,050 Btu. This would indicate that of these three major constituents of coal, sulfur produces the least amount of energy. Therefore, desulfurizing coal reduces the energy produced by a relatively small degree. Furthermore, the removal of sulfur improves the burning of carbon as well as reducing pollution from sulfur oxides.

Combustion may be categorized as furnace combustion, flare combustion, or catalytic combustion.

Furnace combustion features contaminated process gases, auxiliary fuel, combustion air, and a combustion chamber. Furnace combustion is classified as *thermal oxidation* when contaminated gases have a combustible content below the lower explosive limit. Thermal oxidation of gas contaminant into carbon dioxide and water vapor is accomplished by exposure to a fuel- and air-fired flame that raises the incinerator temperature to ranges of 1,000 to 1,500°F and by turbulent mixing. A short intense blue flame permits complete oxidation within a confined space. Figure 10–7 shows this technique.

Furnace combustion is classified as *direct-flame* combustion when contaminated gases contain sufficient combustible to develop a flame

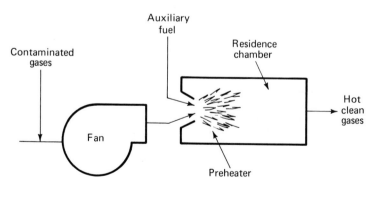

Thermal Oxidation

Fig. 10–7 Thermal oxidation (Courtesy of OMD-APCO-EPA).

in the presence of a proper amount of auxiliary air. The heat generated oxidizes contaminants for a clean discharge to the atmosphere. This type of combustion produces a luminous yellow flame. Figure 10–8 shows this technique.

Furnace combustion may be used to control methyl mercaptan, hydrogen sulfide, and methyl sulfide odors from kraft pulping process; vapor control from paint and varnish cookers; odor control from coffee roasters; and vapor and particulate control from flue-fed apartment-house-sized incinerators. *Afterburners* are used to effect complete combustion of the incinerator effluent gases, and a settling chamber is attached to collect particulates. Figures 10–9 and 10–10 shows an apartment-house incinerator, afterburner, and settling chamber.

Flare combustion, sometimes referred to as direct combustion, is accomplished by directly mixing a gas with air to produce an open flame. A pilot light at the top of the stack is used to initially ignite the flame. The open flame results when oxygen in the air surrounding the flame comes in contact with the hydrocarbons by diffusion only. All process plants that handle hydrocarbons, hydrogen, ammonia, hydrogen cyanide, or other toxic or dangerous gases are subject to emergency conditions which occasionally require immediate release of large volumes of such gases for protection of plant and personnel. Flare combustion is the best means of disposal of these pollutants. Figure 10–11 illustrates a steam-injected flare.

Catalytic combustion is a low-temperature flameless means of gaseous emission control used when gases are cooler and cleaner—that is, when the gas stream contains vaporized or gaseous combustible

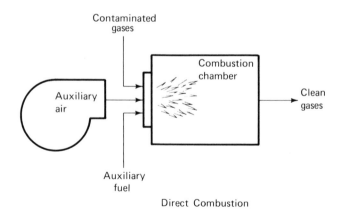

Direct Combustion

Fig. 10–8 Direct combustion (Courtesy of OMD-APCO-EPA).

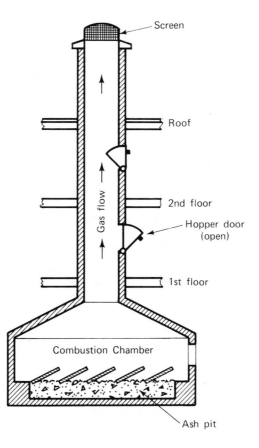

Conventional Single Chamber Flue-Fed Incinerator

Fig. 10–9 Flue-fed incinerator (Courtesy of OMD-APCO-EPA).

materials with no large amount of particulates. The process gases, at a temperature range of 600°F, are directed through or over a bed of catalyst material, wherein temperature is increased and oxidation develops to consume or destroy undesirable elements in the gases. The catalyst accelerates the rate of oxidation of the combustibles in the process gases and in concentrations below their flammable range. This technique requires lower fuel consumption. Platinum alloys, as well as some oxides or vanadium pentoxides, are often used as catalysts because of their ability to produce the lowest catalytic ignition temperatures. Some applications of this technique are lithograph ovens, paint baking ovens, nitric acid manufacturing plants, and oil- and

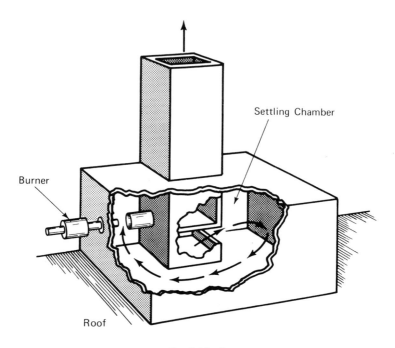

Roof Afterburner

Fig. 10–10 Roof afterburner (Courtesy of OMD-APCO-EPA).

fat-rendering plants. Figure 10–12 illustrates a catalytic combustion system.

other controls of gaseous pollutants

In the case of petroleum storage tanks, where there is considerable evaporation, it is possible to collect and condense the hydrocarbon vapors in a floating roof tank to prevent loss to the atmosphere. This is referred to as a *closed collection and recovery system.*

In the control of odors, *masking and counteracting* may be accomplished by adding an element with a "pleasant" odor in high enough concentration to mask the "malodor" or by combining a pair of odors mixed in appropriate concentrations so that they cancel each other out. Care must be taken to ensure odor counteractants used in work areas are not toxic, allergenic, inflammable, or corrosive. An example of masking is the injection of vanilla into a primary clarifier at a sewage treatment plant to reduce odors caused by the escape of

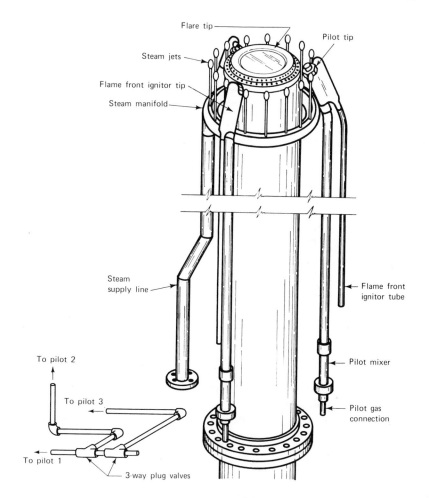

Fig. 10–11 Steam injection type flare (Courtesy of OMD-APCO-EPA).

hydrogen sulfide and methane. Many odorants can exist as liquids under ambient conditions; cooling such vapors can remove much of the odor by simple *vapor condensation.* Rendering-plant cookers and pulp mills are two applications of this technique. Many organic gases and vapors responsible for odors can be converted to odorless compounds by means of relatively gentle forms of *chemical oxidation,* using oxidants such as chlorine, ozone, or potassium permanganate.

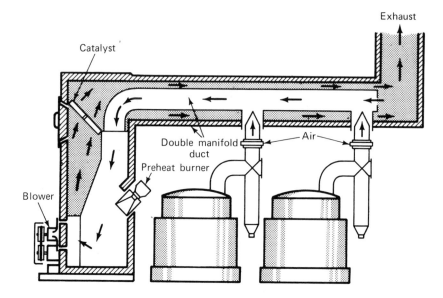

Catalytic Combustion Unit For Waste Fuel Control On A Cooking Kettle Line

Fig. 10–12 Catalytic combustion (Courtesy of OMD-APCO-EPA).

Removal of odors from fish-meal processing is an example of the chemical oxidation application.

A summary of methods for control of specific gaseous pollutants from stationary sources is contained in the control technique manuals listed in the references at the end of this chapter.

DEVICES AND TECHNIQUES FOR CONTROL OF PARTICULATE POLLUTANTS

Particulate material found in ambient air originates from both stationary and mobile sources. (This chapter deals with stationary sources; Chapter 11 deals with mobile sources.) An estimate of particulate emissions into the U.S. atmosphere for 1969 has been broken down as follows:

Millions of Tons

14.4	Industrial processes
7.2	Fuel combustion in stationary sources (mostly power generation)

0.8	Transportation or mobile sources
1.4	Solid-waste disposal
11.4	Miscellaneous

35.2	Total

To control the sources or reduce the effects of particulate pollution, the following techniques can be used:

1/ Gas-cleaning devices for removal of particulates.

2/ Source relocations.

3/ Fuel substitution.

4/ Process changes.

5/ Good operating practices.

6/ Source shutdown.

7/ Dispersion.

All these techniques have been covered previously under control of gaseous pollutants except gas-cleaning devices for removal of particulates, which will be discussed now.

criteria for selection

The selection of a specific type collector or cleaning device involves the following considerations:

Particulate characteristics   The size, shape, density, stickiness, hygroscopicity (tendency to absorb or attract moisture), electrical properties, flammability, corrosiveness, abrasiveness, flowability, and toxicity of particles.

Carrier-gas properties   The temperature, moisture content, corrosiveness, flammability, pressure, humidity, density, viscosity, electrical conductivity, and toxicity of the carrier gas that contains the particulates.

Process factors   Emission rates of specific processes (refer to table 4, Public Health Service Publication No. AP-51), gas flow rates, particle concentration, allowable pressure drop, continuous or intermittent operation, desired efficiency, and ultimate waste disposal.

Economic factors   Refer to tables 5 and 6, PHS Pub. No. AP-51; installation costs, operating costs, and maintenance costs.

Collection efficiency   The particle size gradation in the inlet gas stream is very important in the selection of control equipment since collection efficiency corresponds to particle size and distribution. For instance, large particles can be efficiently collected by a gravitational collector, whereas a fabric filter would be preferable for smaller particles but would tend to clog up rapidly with large particles.

types of gas-cleaning devices

Terminology and classification of these devices are covered under the following major categories: mechanical devices, wet collectors, filter systems, electrostatic precipitators, and afterburners.

Mechanical devices include settling chambers and cyclones.

1/ *Settling chambers* (Fig. 10–13) are closed compartments wherein the velocity of the carrier gas is reduced sufficiently to allow particles of dust and mist to settle out from the gas stream by gravitational force. This device is most efficient for coarse particles (greater than 40 microns in size); the greater the mass, the more efficient. Because of size limitations, it is not economical to use settling chambers for removal of small particles. Settling chambers are often used as *precleaners* to remove large particles, followed by other more efficient devices for removal of smaller

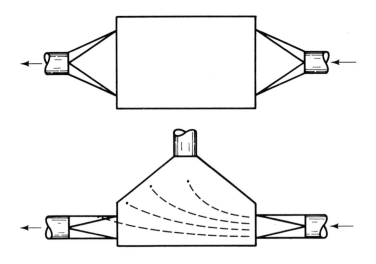

Fig. 10–13 Settling chamber (Courtesy of OMD-APCO-EPA).

particles; this method reduces clogging of small-particle collectors. Some major applications are in kilns, furnaces, and alfalfa feed mills.

2/ A *cyclone* (or dry centrifugal collector) consists of a cylindrical shell fitted with a tangential inlet through which the dust-laden gas enters, an axial exit pipe for discharging the cleaned gas, a conical base, and a hopper to facilitate the collection and removal of dust. Dust-laden gas aided by a fan is swirled in the cylindrical and conical section by admitting it tangentially at the periphery. The gas proceeds downward into the conical section, forms another spiral upward within the downward spiral, and thence travels to the outlet. Particles that are thrown from the rotating streamlines and are able to reach the walls of the cyclone slide down to the hopper.

Large cyclones may be used singly; however, several low-capacity units arranged in parallel may increase efficiency, owing to shortening the streamline radius and thereby increasing the centrifugal force. Small cyclones present some difficulties in that they are more prone to clogging and high-velocity abrasion than a large cyclone.

Fig. 10–14 Multiclone-mechanical dust collector (Courtesy of Western Precipitation Div., Joy Mfg. Co.)

Some major applications of cyclones are in cement factories, metal industries, food and grain mills, asphalt plants, and petroleum refineries. Figure 10–14 shows the multicyclone mechanical dust collector produced by Western Precipitation Division. This is a parallel operation.

Wet collectors can be classified according to method of particle collection as liquid carriage or particle conditioning. *Liquid carriage* consists of a moving body of liquid carrying trapped dust particles to a location outside the collector. Separation of particles from the gas stream occurs when the particles are made to strike a liquid surface within the collector. The liquid serves to prevent particle reentrainment and to carry the particles to some place of ultimate disposal. *Particle conditioning* consists of increasing the effective size of small particles so that collection can be better accomplished by other collection mechanisms. The increase in size may be accomplished by condensation of water upon particles by causing the water temperature to pass through its dew point. Increase in size of small particles may also be induced by the interception of fine dust particles by liquid droplets, resulting in a heavier dust–liquid agglomerate.

Wet collector devices include the following:

1/ In the *gravity spray tower* there are a number of liquid spherical obstacles (droplets) falling in an empty tower by the action of gravity into the path of rising particles, which the droplets collect.

2/ In the *Venturi scrubber,* the dust-laden gas passes through a duct that incorporates a narrow-throated Venturi section into which water spray is injected by radial jets. The atomized droplets collect dust particles from the carrier gas.

3/ A *disintegrator scrubber* consists of an outer casing containing alternate rows of stator and rotor bars. Water is injected axially and is effectively atomized into fine droplets by the rapidly rotating vanes. The droplets collect the dust particles.

4/ *Wet-type dynamic precipitators* combine the dynamic forces of a rotating fan wheel to cause the dust particles to impinge on numerous specially shaped blades. A film is maintained on the blades by spray nozzles.

5/ The *wet impingement scrubber* is a tower consisting of a vertical shell in which are mounted a large number of equally spaced circular perforated (orifice) plates. At one side of each plate a conduit called a downspout is provided to pass the liquid from the plate to the plate below; on the opposite side a similar conduit feeds liquid upward. The dust-laden gas passes through the holes and impinges on the target plates and upon the atomized spray droplets.

6/ *Wet centrifugal scrubbers* rely upon throwing particles against wetted collection surfaces or upon impaction of particles upon droplets and then spraying the dirty droplets upon a wall for collection.

Wet collectors are used widely in kraft paper mills, foundries, fertilizer plants, mining, paint spraying, electroplating, and chemical processing. A critical factor with wet collectors is water usage and waste disposal. Settling tanks and ponds may aid in waste disposal. Filtration and chemical treatment may be used to allow recirculation of water for reuse. Figure 10–15 shows the Turbulaire wet scrubber produced by Western Precipitation Division.

Filter systems consist of a woven or felted fabric through which dust-laden gases are forced. These fabrics are woven into bags and enclosed in a *bag house*. A combination of factors results in the collection of particles on the fabric fibers. When woven fabrics are used, a dust cake eventually forms, which in turn acts predominantly as a sieving mechanism. When felted fabrics are used this dust cake is

Fig. 10–15 Turbulaire scrubber (Courtesy of Western-Precipitation Div. Joy Mfg. Co.)

minimal or nonexistent due to the nature of the fabric. Instead, the main filtering mechanisms are a combination of inertial forces, electrostatic forces, impingement, diffusion, and gravitational settling as related to individual particle collection on single fibers.

As particulates are collected on the fabric, pressure drop across the filtering media increases. Filters must then be cleaned in order not to overload the fans that are forcing the dust-laden gases through the bags. The cleaning may be done by shaking the bags to remove dust for disposal, or it is removed by rapping, sonic waves from an air horn, reversing the airflow, pressure jets, or gentle collapsing.

Various types of fabrics and different weave patterns are used in bag houses. Some factors to be considered in selecting fabrics are their melting temperature, acid or alkali resistance, air permeability, and resistance to abrasion and shrinkage. Some fibers that are used include cotton, wool, nylon, Nomex, asbestos, Orlon, Dacron, silicon-treated woven glass, and Teflon.

Bag houses are often placed in series after a mechanical collector to capture small particulates not captured by the mechanical collector. Since larger particles have already been removed, bag wear and frequency of cleaning are reduced. Particles down to 0.01 μ in size can be

Fig. 10–16 Therm-O-flex filter   (Courtesy of Western-Precipitation Div. Joy Mfg. Co.)

Fig. 10–17 Electrical precipitator (Courtesy of Western-Precipitation Div. Joy Mfg. Co.)

efficiently collected. Some applications are cement kilns, open-hearth furnaces, steel furnaces, and grain-handling operations. Figure 10–16 shows the Thermoflex filter produced by Western Precipitation Division.

Electrostatic precipitators utilize extremely high voltage electric current to separate dust, fumes, or mist from a gas stream. Four basic steps are involved: electrically charging the particles by ionization, transporting the charged particles to a collecting surface by the force exerted upon them in the electric field, neutralizing the electrically charged particles precipitated on the collecting surface, and removing the precipitated particles from the collecting surface. Removal of particles is performed by rapping or by washing.

These devices can collect very small particles (1 to 44 μ) with high efficiency. Where excessive dust loads occur, a mechanical collector may be placed in series before the electrostatic precipitator. Acids, high-temperature wastes, and corrosive material that might damage a bag house can be collected. Electrostatic precipitators are used in pulverized coal-fired power plants, steel plants, cement kilns, and pulp and paper mills. Figure 10–17 shows an electrical precipitator produced by Western Precipitation Division.

TABLE 10-1   Advantages and Disadvantages of
Particulate Collection Devices

Collector	Advantages	Disadvantages
Gravitational	Low pressure loss, simplicity of design and maintenance.	Much space required. Low collection efficiency.
Cyclone	Simplicity of design and maintenance; little floor space required; dry continuous disposal of collected dusts. Low to moderate pressure loss. Handles large particles. Handdles high dust loadings. Temperature independent.	Much head room required. Low collection efficiency of small particles. Sensitive to variable dust loadings and flow rates. Tends to clog up.
Wet collectors	Simultaneous gas absorption and particle removal. Ability to cool and clean high-temperature, moisture-laden gases. Corrosive gases and mists can be recovered and neutralized. Reduced dust explosion risk. Efficiency can be varied.	Corrosion, erosion problems. Added cost of wastewater treatment and reclamation. Low efficiency on submicron particles. Contamination of effluent stream by liquid entrainment. Freezing problems in cold weather. Reduction in buoyancy and plume rise; water vapor contributes to visible plume under some atmospheric conditions.
Electrostatic precipitator	99+% efficiency obtainable. Very small particles can be collected. Particles may be collected wet or dry. Pressure drops and power requirements are small compared to other high-efficiency collectors. Maintenance is nominal unless corrosive or adhesive materials are handled. Few moving parts. Can be operated at high temperatures ($550°$ to $850°F$).	Relatively high initial cost. Precipitators are sensitive to variable dust loadings or flow rates. Resistivity causes some material to be economically uncollectable. Precautions are required to safeguard personnel from high voltage. Collection efficiencies can deteriorate gradually and imperceptibly.

TABLE 10-1   (Cont'd)

Collector	Advantages	Disadvantages
Fabric filtration	Dry collection possible. Decrease of performance is noticeable. Collection of small particles possible. High efficiencies possible.	Sensitivity to filtering velocity. High-temperature gases must be cooled to $200°F$ to $550°F$. Affected by relative humidity (condensation). Susceptibility of fabric to chemical attack.
Afterburner, direct flame	High removal efficiency of submicron odor-causing particulate matter. Simultaneous disposal of combustible gaseous and particulate matter. Direct disposal of non-toxic gases and wastes to the atmosphere after combustion. Possible heat recovery. Relatively small space requirement. Simple construction. Low maintenance.	High operational cost. Fire hazard. Removes only combustibles.
Afterburner, catalytic	Same as direct flame afterburner. Compared to direct flame: reduced fuel requirements, reduced temperature, insulation requirements, and fire hazard.	High initial cost. Catalysts subject to poisoning. Catalysts require reactivation.

Extracted from Control Techniques for Particulate Pollutants, AP-51.

Afterburners (direct flame or catalytic) are considered mostly as a control device for gaseous pollutants, but may also be considered for control of particulate residue-free vapors, mists, smoke, and readily combustible particles. They are often used for fume, vapor, and odor control where relatively small volumes of gases low in particulate concentration are involved (see Fig. 10–10).

Advantages and disadvantages of particulate collection devices are listed in Table 10–1.

REFERENCES

U.S. Department of Health, Education, and Welfare, *Air Pollution Control Technique Manuals*, PHS (NAPCA), Pub. Nos.
    AP-51, Particulates, 1969.
    AP-52, Sulfur Oxide, 1969.
    AP-65, Carbon Monoxide from Stationary Sources, 1970.
    AP-67, Nitrogen Oxide from Stationary Sources, 1970.
    AP-68, Hydrocarbons and Organic Solvent from Stationary Sources, 1970.
Washington, D.C.: U.S. Government Printing Office.

U.S. Department of Health, Education, and Welfare, *Air Pollution Engineer Manual*, PHS, Pub. No. 999-AP-40. Washington, D.C.: U.S. Government Printing Office, 1967.

RECOMMENDED FILMS

MA-47   Clean Combustion (30 min)

MA-39   Collection of Particulates and Control of Air Pollution

MA-57   Combustion for Control of Gaseous Pollutants (20 min)

MA-14   Control of Air Pollution (5 min)

MA-30   Principles of combustion (20 min)

TF-104   Particulate Control Device—Cyclone (3 min)

TF-105   Particulate Control Device—Electronic Precipitator (3 min)

TF-107   Particulate Control Device—Fabric Filters (3 min)

TF-103   Particulate Control Device—Settling Chamber (3 min)

TF-106   Particulate Control Device—Venturi Scrubber (3 min)
    Available: Distribution Branch
               National Audio-Visual Center (GSA)
               Washington, D.C. 20409

QUESTIONS

1/ What is the most positive way to abate air pollution?

2/ What are the two fundamental approaches that can be taken to reduce air pollution?

3/ How is effective stack height estimated?

4/ What are the five gaseous pollutant emission control methods? Give examples of each.

5/ What four elements are necessary in proper proportions to obtain complete combustion?

6/ What is the difference between thermal oxidation and direct combustion?

7/ When is catalytic combustion preferred over other forms of combustion?

8/ What gaseous pollutant emission control principle is used to prevent odors from a primary clarifier in a sewage treatment plant?

9/ What are the five major categories of gas-cleaning devices used for removal of particulates? Give an example of each.

10/ Which pollutant collector is most efficient in capturing very small particulates of a corrosive nature?

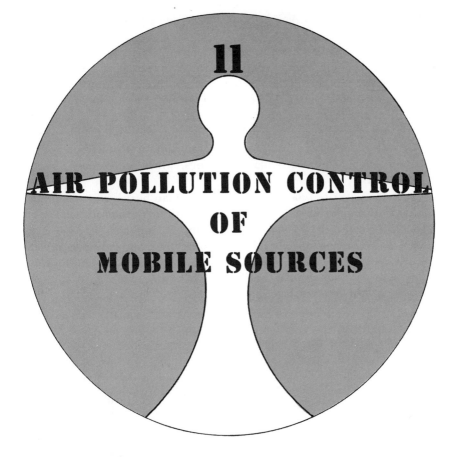

11

AIR POLLUTION CONTROL
OF
MOBILE SOURCES

Student Objectives

—*To develop an awareness of the air pollutants emitted by mobile sources and what is being done to reduce these emissions*

—*To know the role of the federal government and the state governments regarding control of mobile source emissions and fuel standards*

—*To become familiar with mobile source emission standards and the various devices and techniques applied in an attempt to reach these standards*

AUTOMOBILE POLLUTANTS OF CONCERN

In his State of the Union address in January, 1970, President Nixon stated that the automobile was our worst pollution source. In

1970 HEW published data based on the year 1968 (shown on a bar graph, Fig. 11–1). This graph indicates that transportation contributed about 42 per cent of the major pollutants by weight. Motor vehicles alone contributed about 39 percent. Transportation contributed 50 percent of the man-made hydrocarbons (HC), a little over 60 percent of the total carbon monoxide (CO) emissions, and about 40 percent of the nitrous oxides (NO_x). However, transportation contributes very little of the SO_x and particulate emissions, which come mostly from fixed sources.

If the data are related to effects on human health and plant life rather than weight of pollutants, transportation is responsible for less than 10 percent of the total U.S. air pollution problem for 1968 (see Fig. 11–2). For example, it takes 200 tons of CO to be of as much health concern as 1 ton of SO_x, according to regulatory authorities quoted by William Agnew in *Progress in Areas of Public Concern*, General Motors Proving Ground Report.

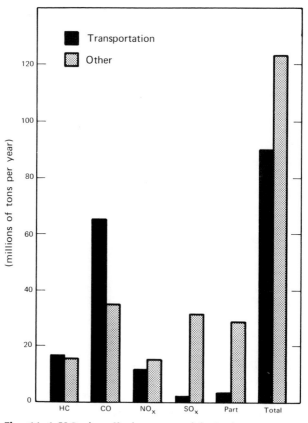

Fig. 11–1 U.S. air pollution on a weight basis.

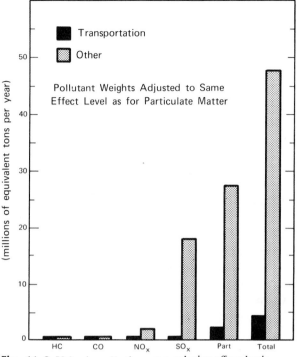

Fig. 11–2 U.S. air pollution on a relative effect basis.

Data from the EPA for 1969 national emissions show a total of 281.2 million tons with 144.4 from transportation (51 percent). Transportation contributed 74 percent of the CO, 53 percent of the HC, 50 percent of the NO_x, and only 0.03 percent of the SO_x and 0.02 percent of the particulates emitted that year.

Data for uncontrolled automobiles indicate that for every 1,000 gal of gas consumed, the following amounts of pollutants are produced: 2,300 lb of CO, 200 lb of HC, 113 lb of NO_x, 12 lb of particulates, 9 lb of SO_x, 4 lb of organic acetic acid, and 4 lb of aldehydes. From these uncontrolled levels, the *reduction* resulting from 1971 model control devices is estimated to be 80 percent HC, 69 percent CO, and in California, where NO_x controls are mandatory, 33 percent NO_x. This would mean 713 lb of CO, 40 lb of HC, and, in California, 76 lb of NO_x as *expected emission factors* per 1,000 gal of gasoline burned in 1971 models. This is a major reduction, but not as good as the EPA will require in 1975–1976 models.

Of the three primary gaseous pollutants emitted directly from the auto exhaust (CO, HC, and NO_x), the most toxic is CO; however, HC and NO_x are also of grave concern. Secondary pollutants or harmful smog products formed after the primary pollutants reach the atmosphere are also of grave concern. These are NO_2, O_3, PAN, and aldehydes that have been discussed in previous chapters. Particulates of most concern are lead and carbon. It is hoped that nonleaded gasoline now being developed will reduce the hazard of lead in the atmosphere coming from automobiles. Other air pollutants being investigated include barium, which is used as a smoke suppresser; fuel additives such as boron, manganese, and nickel; phosphorus added as a corrosion inhibitor; cadmium and hydrochloric acid from tetraethyllead gas; asbestos; and HC suspected as organic carcinogens.

SOURCES OF POLLUTANTS FROM AUTOMOBILES

The auto crankcase and tailpipe exhaust of *uncontrolled* vehicles contribute 85 percent of the air pollution problem. From the crankcase, unchanged lubricant oil, CO, and HC emanate. From the tailpipe, gaseous ˇexhaust, including CO, HC, C, NO–NO_2 and lead, is expelled. Evaporation from the carburetor and fuel tank, consisting mainly of HC, account for the remaining 15 percent. Of the total HC, it is estimated that 20 percent comes from the crankcase, 20 percent from evaporation, and 60 percent from the exhaust. Of the total CO, most comes from the exhaust. All the NO_x and lead are expelled from the exhaust.

LEGISLATION AFFECTING MOTOR VEHICLE EMISSIONS AND FUEL STANDARDS

the federal role

The *federal role* in controlling auto emissions has now been well established. The first legislation to control auto emissions was passed in 1947, when California prescribed the first crankcase controls. The last vehicles to be produced without controls for California were 1961 models, and for the nation as a whole, the 1962 models. In 1963 U.S. auto manufacturers voluntarily installed a blow-by control device to eliminate crankcase emissions on all models for nationwide use. In 1968 the federal government established the first national exhaust emission standards; and in the 1970 amendments to the Clean Air Act, standards were established for 1973–1974 models and projected to 1975–1976 models./ Proposals reflected through December 1972 are included in Table 11–1.

TABLE 11-1   Auto Air Pollution Emission Standards(a)

	1968	1970	1971	1973-74	1975-76
Crankcase	none	- -	- -	- -	- -
Exhaust(b)					
HC	275 ppm	2.2g/mi(c)	2.2g/mi	3.4g/mi	0.41g/mi
CO	1.5%	23g/mi(c)	23g/mi	3.9g/mi	3.4g/mi
NO_x (as NO_2)	- -	- -	- -	3.0g/mi	0.40g/mi
Evaporation (HC)	- -	- -	6g/test(d)	2g/test	2g/test
Particulates(e)	- -	- -	- -	- -	0.1g/test

(a) Diesel fueled light duty vehicles emission standards for HC, CO, and NO_x are same as for gas fueled vehicles. No evaporation standards have been established.

(b) Heavy duty engines, gasoline and diesel emission standards for 1973-74 are HC plus NO_x 16 grams per brake horse power hour and 40 grams CO per brake HP hour.

(c) For light duty vehicles. Engine displacement of less than 50 cubic inches. For heavy duty gasoline engines, HC 275 ppm, CO 1.5% by volume.

(d) Does not apply to off-the-road utility vehicles until 1972.

(e) Federal tests for diesels in 1973 have been set at 20% opacity during acceleration and 15% opacity during lugging, 50% at peak. (20% opacity is equivalent to Ringelmann No. 1, 40% equivalent to Ringelmann No. 2, and 50% equivalent to No. 2.5.)

The Clean Air Act establishes the right of the federal government to set *emission standards* for all *new* vehicles and to test these vehicles to ensure compliance. The auto manufacturer must maintain records available to the EPA for inspection to ensure compliance, and each vehicle must be certified as complying with established standards. Figure 11–3 shows the required certificate as it appears on Chrysler automobiles. This prescribes that the manufacturer or dealer must not remove or render air pollution devices inoperative before or after sale to the ultimate buyer. The manufacturer must also furnish proper instructions for maintenance of vehicles to retain compliance with air pollution regulations. Anyone violating one of these requirements is subject to a $10,000 fine. Acting for the federal government, the EPA has required a reduction of 90 percent in HC and CO emitted by 1975 models as compared with 1970 requirements, and a 90 percent reduction in oxides of nitrogen by 1976. The federal government through the EPA also must register fuel or fuel additives designed to reduce air pollution before they can be sold. Violation of this requirement may result in a $10,000/day fine for the producer of the

Fig. 11–3 Vehicle emission control certificate (Courtesy of Chrysler Corp.).

unregistered fuel. The EPA may regulate or prohibit the manufacture or sale of fuels or fuel additives that result in harmful emissions or interfere with motor vehicle pollution control devices. The EPA has determined that lead exceeding 2 μg/m³ average over a period of 3 months or longer may endanger public health. A reduction of 60 to 65 percent lead additives to fuel will be required. The EPA is also considering the lowering of sulfur content of vehicle fuels, as well as lowering phosphorus content.

The EPA has proposed limiting gasoline additives by 1974 to 2.0 g of lead/gal and 0.01 g of phosphorus/gal. Further proposals are that "lead-free" gas will contain no more than 0.05 g of lead/gal. "Phosphorus-free" gas will contain no more than 0.01 g of phosphorus/gal. Under new federal regulations, gas stations must carry one grade of unleaded gasoline by July 1, 1974. This is to protect catalytic devices needed in 1975–1976 models for reduction of NO_x emissions.

the state role

The *state government role* includes responsibility for controls, regulations, restrictions, and testing of all vehicles after sale to the ultimate purchaser. Some approaches to ensure control are

1/ A check can be made to determine the presence of required control systems as part of a state's safety inspection.

   a/ Place the vehicle on a dynamometer to determine if emissions from the exhaust meet emission standards.

   b/ Use vacuum check to ensure that the crankcase control device is operable.

   c/ Inspect to see if evaporation control equipment is installed.

2/ Inspect standing or moving vehicles for excessive visual smoke and issue fine or warning to obtain necessary repairs or modifications within 30 days.

3/ Require a state inspection of *all* cars prior to sale to ensure that vehicles comply with air pollution control standards.

4/ Require installation of a crankcase control device and exhaust emission kit on all uncontrolled vehicles (a General Motors kit is available for approximately $20).

5/ Push for substitution of public mass transportation systems to replace some private vehicles.

6/ Design or redesign roads and *traffic-control* patterns to reduce traffic density or stagnation. Examples:

   a/ Oneway streets.

   b/ Express lanes.

   c/ Graduated tolls based on number of passengers.

   d/ Ban on-street parking.

   e/ Stagger work hours.

7/ Stress better driving methods and better maintenance to reduce air pollution as part of driver-training education programs.

8/ Certify maintenance personnel for operation of air pollution control check systems included with other maintenance checks.

9/ Encourage use of other vehicular propulsion systems to replace the gasoline engine.

10/ Establish emergency plans to reduce vehicle traffic when high levels of pollution threaten an air pollution episode.

11/ Require fleets to convert to fuels other than gasoline.

AIR POLLUTION CONTROL DEVICES FOR AUTOMOBILES

control devices for crankcase emissions

Crankcase emissions or blow-by (leakage between piston ring and cylinder wall) have been virtually eliminated nationwide on all cars produced since 1962. A control device called positive crankcase ventilation (PCV) prevents blow-by from the crankcase by returning gases to the cylinders to be burned inside the engine instead of being vented into the atmosphere. To function properly, the PCV or smog valve must be clean and should be inspected at every oil change and be re-

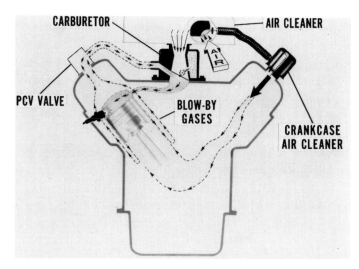

Fig. 11–4 Closed crankcase ventilation system (Courtesy of Chrysler Corp.).

placed every 12,000 miles or 12 months, whichever comes first. Figure 11–4 shows a close-up view of the closed crankcase ventilation system used by Chrysler.

control devices for exhaust emissions

Exhaust emissions are reduced in various ways. One approach is based on the principle of reducing HC and CO emissions by adding fresh air to the hot exhaust gas to supply necessary oxygen needed for more complete burning as the mixture moves through the exhaust system. This system also includes engine modifications to increase the effectiveness of burning within the combustion chamber itself. Modifications include an engine-driven air pump, air control and distribution equipment to deliver air to each exhaust port, carburetor modifications, distributor and vacuum advance modification, and manufacturers recommended maintenance.

Another approach is the controlled combustion system, which utilizes engine design parameters to achieve emission control through more thorough combustion. This system includes an air–fuel mixture control, modified ignition timing, and recommended maintenance. Dupont has designed an exhaust manifold thermal reactor for reducing HC and CO emissions that can be installed on current vehicles. Exhaust emission device systems require periodic servicing of ignition and carburetor systems, especially engine idle speeds, spark timing and idle fuel–air ratio adjustments.

control devices for evaporation emissions

Fuel evaporation emissions have been reduced by the installation of a fuel tank with a built-in chamber to provide an assured thermal expansion volume for the fuel. The tank is vented into a vapor–liquid separator, which returns liquid to the tank and passes vapor through a pressure-vacuum relief valve to an activated carbon canister. Vapor is stored in the canister until purged by a vacuum created when engine operation empties the vapor into the intake manifold and then into the combustion chamber, where it is burned. The air filter in the base of the carbon canister should be replaced every 12 months or 12,000 miles. Some vapor control systems have no carbon canister, but instead use the engine crankcase as a storage container.

The General Motors 1971 emission control system is shown in Fig. 11–5. This system permitted vehicles to meet the 2.2 g of HC and 23 g of CO per vehicle mile restrictions of 1971. The evaporation control system controlled evaporation losses to the 6 g/test limit of that year.

more control devices needed

The federal government's emission standards for NO_x (which became effective in 1973–1974 models) require some additional controls. Dupont and Esso have been researching a device based on the principle that recirculation of a portion of the exhaust dilutes the fuel–air mixture and effectively reduces the formation of NO_x. Gases are taken from the exhaust pipe just ahead of the muffler and directed into the carburetor between the Venturi section and the throttle plate.

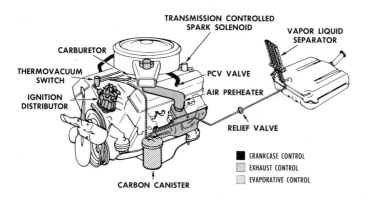

Fig. 11–5 General Motors 1971 emission control system (Courtesy of General Motors Corp.).

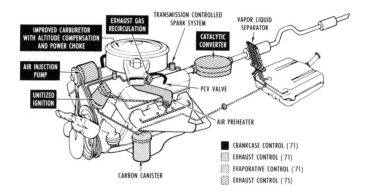

IMPROVED CARBURETOR
WITH ALTITUDE COMPENSATION
AND POWER CHOKE

EXHAUST GAS
RECIRCULATION

TRANSMISSION CONTROLLED
SPARK SYSTEM

VAPOR LIQUID
SEPARATOR

CATALYTIC
CONVERTER

AIR INJECTION
PUMP

UNITIZED
IGNITION

PCV VALVE

AIR PREHEATER

CARBON CANISTER

CRANKCASE CONTROL ('71)
EXHAUST CONTROL ('71)
EVAPORATIVE CONTROL ('71)
EXHAUST CONTROL ('75)

Fig. 11–6 General Motors 1975 emission control system for lead free fuel
(Courtesy of General Motors Corp.).

Figure 11–6 shows the General Motors 1975 emission control system, which is basically the same as for 1971 with some differences, that is, the unitized ignition, the air injection pump, and the catalytic converter designed to reduce NO_x emissions to meet 1975 emission standards. This system depends upon the use of lead-free fuel because lead would foul the catalytic converter.

All model 1971 automobiles for sale in California were required to have a control system limiting NO_x emissions. Chrysler's NO_x control system is shown in Fig. 11–7. Control of NO_x is achieved by lowering peak burning temperatures during combustion. In Chrysler's Cleaner Air System, control is accomplished by a combination of factors involving valve timing, retarding spark advance at low vehicle speeds, and using lower temperature thermostats. Use of a modified camshaft causes an increase in valve overlap with the result that both the intake and exhaust valves are open longer at the same time. During this interval, the incoming fuel–air charge to be burned is diluted by exhaust gases being discharged from the cylinder. This slight dilution is very important in controlling peak combustion temperatures. Spark advance is controlled by the solenoid switch located between the distributor and carburetor. The solenoid, in turn, is controlled by three switches that sense car speed, air temperature, and manifold vacuum. No spark advance is allowed by the solenoid on accelerations up to 30 mph if the manifold vacuum is less than 15 in. mercury. Spark advance is restored, however, at temperatures below approximately 60°F. The operating temperature of cooling systems has been lowered by use of 185°F thermostats. Lower coolant temperatures assist in controlling temperatures in the cylinders.

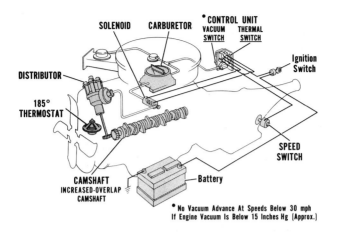

Fig. 11–7 NO$_x$ control system, automatic transmission (Courtesy of Chrysler Corp.).

Ford Motor Company's low emission concept vehicles, which incorporate the use of dual bed catalytic converters for conversion of HC and CO as well as NO$_x$ and the use of nonleaded gasoline, have made significant strides toward attaining the future federal emission standards listed in Table 11–1.

The next target for reduction by the federal government is *particulates*. It is estimated that new automobiles are now emitting 0.3 g/mile; the 1975 target is 0.1 with 0.03 proposed for 1980. The removal of lead from gasoline will help reduce particulates. Not only lead itself, but cadmium and hydrochloric acid, which are used in production of leaded gasoline, are of concern. Removal of lead will also make it possible to use the catalytic converter for removal of NO$_x$. Dupont has experimented with a system that separates particulates from the exhaust gas stream by trapping them with a cyclone separator. Exhaust gas is cooled during passage through the exhaust pipe so that particles become solids. The exhaust stream is passed through a box filled with wire mesh to cause the particles to collide and agglomerate into larger particles. The agglomerated particles are then removed from the stream in a cyclone separator and retained in a collection box for periodic removal.

alternatives to the gasoline engine

Many alternatives to the gasoline engine have been suggested, including the steam engine, the gas turbine engine (utilizes an excess of air that reduces HC and CO, but does not control NO emissions), the stratified charge-fuel injection engine, the free-piston diesel engine (more smoke and smell and more NO_x and SO_x, but less HC and CO), the electric car (zinc–air battery, lithium–nickel–halide battery, sodium–sulfur battery), the Wankle engine (rotary combustion chamber), and the Stirling engine (thermodynamic-external combustion-air engine). Of these, the one appearing to have best possibility is an electric car using a fuel cell (a special form of battery refuelable like a gasoline engine), which uses hydrazine, ammonia, or alcohol linked with an auxiliary fast-discharge battery for peak acceleration.

Improved fuels are another approach; these include diesel fuel, liquid petroleum gas (LPG), compressed natural gas (CNG), liquid natural gas (LNG), solar energy, nuclear power, and recently a fuel produced from liquid hydrogen. In England a farmer runs his car on methane gas produced from chicken manure. In a turbine engine, peanut oil or almost anything burnable can be used for fuel.

Mass transportation systems may be the ultimate solution to traffic problems in the big cities, where traffic and pollution are becoming extremely critical problems.

DIESEL-POWERED VEHICLES

The large trucks burning diesel fuel are a major contributor to air pollution problems, particularly in the emissions of particulates (C and HC aerosols), visible smoke, and irritating odors. Diesel trucks produce more NO_x and SO_x than gasoline-burning vehicles. A comparison of various pollutants is shown in Table 11–2. Footnotes with Table 11–1 list emission standards for diesels.

TABLE 11-2   Auto vs. Diesel Emission Factors
(Pounds per 1000 Gal. of Fuel)

Automobiles		*Diesel Trucks*
4	Aldehydes	10
2300	CO	60
200	HC	136
113	NO_x	222
9	SO_x	40
4	Acetic acid	31
12	Particulates	110

No controls have been developed for diesel exhaust emissions; however, proper maintenance can reduce the emissions considerably, especially those found in the visible black smoke. Barium-containing fuel additives have been used to reduce black smoke. Isopropyl nitrate has been used to suppress white smoke. Crankcase and evaporative emissions are extremely low.

Some states have regulations limiting the density of diesel smoke, which must not exceed Ringelmann no. 1, and use either a smokemeter for reading the smoke or the trained eye of a smoke reader. One state uses the smokemeter and the dynamometer to check all trucks at truck weighing stations.

AIRCRAFT

The Clean Air Act Amendments of 1970 directed the EPA to make a study and investigation of emissions of air pollutants from aircraft. A report was published by the EPA in 1972 presenting the available information on present and predicted nature and extent of air pollution related to aircraft operations in the United States and the technological feasibility of controlling such emissions. This study led to the conclusions that aircraft emissions are significant contributors to air pollution, particularly in the vicinity of an airport, and must be controlled if national ambient air quality standards are to be met in those areas. The study also concluded that emissions could be reduced by modification of ground operational procedures, improvement in maintenance and quality control procedures from existing aircraft, development and demonstration of new combustion technology for new aircraft, and retrofit of turbine engine fleets with existing technology for near-term reduction of emissions.

modification to ground operational procedures

Although no regulations have been proposed in Federal Register Vol. 37, No. 239, pp. 26502–26503, dated December 12, 1972, regarding modification to ground operation of aircraft to control emissions, this Register expressed the views of the EPA that more immediate measures could be employed in this area. The EPA estimated that aircraft emissions of HC and CO at airports could be reduced by 50 to 70 percent by establishing rules pertaining to idle and taxi modes of ground operations.

The Secretary of Transportation questioned the effects of modification of aircraft ground operations to reduce air pollution as they might affect safety and possibly increase noise pollution. Because of these questions raised, air pollution regulations related to aircraft ground operations are still under study.

proposed aircraft emission standards

In December, 1972, the EPA proposed that new and in-use aircraft gas turbine engines not be allowed to vent fuel beginning in 1974 and that crankcase emissions be eliminated for new aircraft piston engines by 1979.

Proposed exhaust emission standards for new and in-use aircraft gas turbine engines were as follows:

—By 1974 maximum allowable smoke, no. 30 as measured by a reflectometer.

—By 1976, varying by size of engines, HC, 8.7 to 2.5 lb/1,000 lb-thrust h/cycle; CO, 11.3 to 11.9 lb/1,000 lb-thrust h/cycle; smoke ranging from no. 35 to none.

—By 1979, considerable reduction of HC, CO, and smoke over 1976 proposals and adding NO_x emission standards.

Proposed exhaust emission standards for new aircraft piston engines:

—By 1979:   HC, 0.00190 lb/rated power/cycle;
             CO, 0.042   lb/rated power/cycle;
             NO_x, 0.0015 lb/rated power/cycle;

Proposed exhaust emission standards for new gas turbine aircraft:

—By 1976:   HC, 0.4 lb/1,000 hp-h of power output;
             CO, 12.2 lb/1,000 hp-h of power output.

—By 1979:   HC 0.4 lb/1,000 hp-h of power output;
             CO, 5 lb/1,000 hp-h of power output;
             NO_x 3 lb/1,000 hp-h of power output.

The proposals also included test procedures applicable to the determination of these data.

In reference to reaction to an earlier study made in 1968–1969 regarding aircraft emissions, most of the major airlines agreed to a schedule of retrofitting JT8D (Class T4) aircraft engines with reduced smoke combusters. This retrofit program was 85 percent complete by July, 1972, and is a major step toward reduction of this air pollutant from large aircraft.

Many emission inventories do not include air pollution from aircraft; however, near major airports this has become a serious problem. The emissions from aircraft have been figured by pounds per flight (including takeoff and landing) below 3,500 ft (see Table 9–14 for emission factors for aircraft). CO, HC, and NO_2 have been considered pollutants of main concern, but more recently attention has also focused on visible smoke and its particulate content as well as its gaseous pollutant content.

The U.S. Secretary of Transportation will be responsible for enforcing aircraft emission standards whenever they are established. The Federal Aviation Act sets forth aviation fuel standards. States are prohibited from establishing air pollution controls over aircraft.

REFERENCES

General Motors Proving Ground, *Progress in Areas of Public Concern.* Milford, Mich.: General Motors Corp., Feb. 1971.

Society of Automotive Engineers, *The 1970 GM Emission Control Systems,* Pub. No. 700149. Detroit, Mich., Jan. 1970.

U.S. Department of Health, Education, and Welfare, Air Pollution *Control Techniques Manuals* (Carbon Monoxide, Nitrogen Oxide, and Hydrocarbons from Mobile Sources), PHS (NAPCA), Pub. No. AP-66. Washington, D.C.: U.S. Government Printing Office, 1970.

U.S. Environmental Protection Agency, *Aircraft Emissions: Impact on Air Quality and Feasibility of Control.* Washington, D.C.: U.S. Government Printing Office, 1972.

U.S. Environmental Protection Agency, *The Clean Air Act.* Washington, D.C.: U.S. Government Printing Office, Dec. 1970.

U.S. Environmental Protection Agency, *Federal Register,*
Motor Vehicles, Vol. 36, No. 128, pp. 12652, July 2, 1971.
Motor Vehicles, Vol. 37, No. 10, pp. 669–675, Jan. 15, 1972.
Motor Vehicles, Vol. 37, No. 12, pp. 784–785, Jan. 19, 1972.
Motor Vehicles, Vol. 37, No. 24, pp. 2732–2737, Feb. 4, 1972.
Auto Fuels, Vol .37, No. 36, pp. 3881–3884, Feb. 23, 1972.
Auto Fuels, Vol. 37, No. 50, pp. 5303–5304, Mar. 14, 1972.
Heavy Duty Engines, Vol. 37, No. 175, pp. 18262–18270, Sept. 8, 1972.
Motor Vehicles, Vol. 37, No. 193, pp. 20914–20923, Oct. 4, 1972.
Motor Vehicles, Vol. 37, No. 221, pp. 24250–24320, Nov. 15, 1972.
Aircraft, Vol. 37, No. 239, pp. 26488–26503, Dec. 12, 1972.
Washington, D.C.: U.S. Government Printing Office

RECOMMENDED FILMS

Air Pollution and Cars (20 min)

Answer Is Clear (14 min)

Combustion in Action (19 min)
        Available: General Motors Film Library
                     Detroit, Mich. 48202

MA-16   No Smoking (7 min)

MA-82   Vehicle Emissions Control Story (25 min)

M-1581   Cleaner Cars for Cleaner Air (15 min)
        Available: Distribution Branch
                     National Audio-Visual Center (GSA)
                     Washington, D.C. 20409

QUESTIONS

1/ What was considered the worst source of air pollution in 1970?

2/ What are the three main sources of pollutants from uncontrolled automobiles?

3/ Who is responsible for establishing motor vehicle emission and fuel standards for new vehicles?

4/ Who is responsible for air pollution controls of vehicles after sale to ultimate purchaser?

5/ What pollutants are presently covered by emission standards?

6/ List five options open to state governments for control of air pollution from automobiles.

7/ What is the difference between crankcase emissions and exhaust emissions?

8/ What major pollutant will be included for the first time in national automotive emission standards beginning with model year 1973?

9/ Are there any federal emission standards for diesel-fueled vehicles?

10/ Who has air pollution control over aircraft?

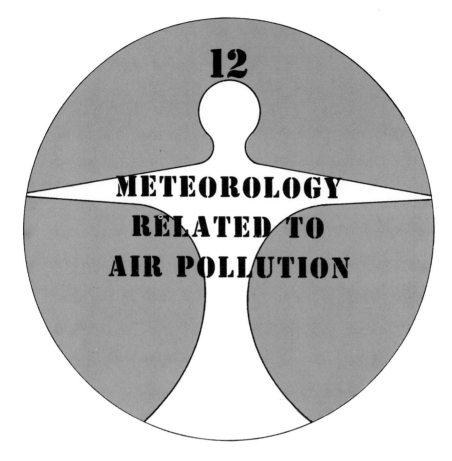

12

METEOROLOGY RELATED TO AIR POLLUTION

Student Objectives

—*To know the primary effects of meteorology on air pollution.*
—*To know the role of meteorology in the transport and diffusion of air pollutants.*
—*To become familiar with the primary meteorological instruments used in air pollution work.*
—*To know the important sources of meteorological information and how assistance on meteorological problems in air pollution may be obtained.*

Air pollution is a problem because human activities threaten to overload the atmosphere with wastes, beyond the ability of wind and weather to disperse and dilute these pollutants. An effective air re-

223

sources management program must take into account the effects of meteorological parameters on transport and diffusion and the natural cleansing processes of the atmosphere.

Particulate pollutants tend to coagulate, increase in size, and fall to earth; thus coarse particulates generally cause air pollution problems of a localized nature. However, lighter particulates and gaseous pollutants are influenced by the action of atmospheric diffusion and may be carried great distances from their source of origin.

This chapter covers basic meteorological fundamentals, effects of meteorological parameters on transport and diffusion, natural cleansing processes in the atmosphere, pollutant concentration variations related to resultant plumes, effects of air pollution on atmospheric visibility, meteorological instruments used to collect meteorological data, and some of the uses of the meteorological data collected.

BASIC FUNDAMENTALS OF METEOROLOGY

Earth's weather is created by four primary interacting elements —the sun, the earth itself, the earth's atmosphere, and natural landforms and geophysical features of the earth's surface. Meteorology is the science of the atmosphere and the study of those elements which are characteristic of weather. Another factor related to meteorology is the effect of air pollution on atmospheric visibility. Because of its importance, visibility will be covered in considerable depth toward the end of this chapter.

radiation from the sun

The sun beams energy through space to earth as electromagnetic waves. Much of this radiation is absorbed by the water vapor in the atmosphere, and it is this heat energy that powers the complex circulation of the atmosphere. Part of the heat energy that reaches the earth is reradiated back into the atmosphere, and much of this is absorbed by water vapor and carbon dioxide, creating the "greenhouse" effect that protects the earth from scorching in the daytime and also conserves heat energy to prevent extensive cold at night.

earth's shape and movement

The roughly spherical shape of the earth combined with the tilt of the earth allows the sun's rays to strike with greater intensity at the equator than at the poles. As a result the equatorial area of the earth would continue to heat up and the poles would continue to cool unless some force allowed for a transfer of heat from the equator to the poles.

As the earth swings around the sun in its elliptical orbit, it also rotates on its axis from west to east. This rotation along with the resultant Coriolis force (a deflecting force acting on a body in motion due to the earth's rotation) tends to deflect winds in the northern hemisphere to the right. The warm air tends to flow from the tropics to the poles more westerly (toward the east), and the cold air from the poles tends to move toward the equator more easterly (toward the west). The result is that most of the atmospheric motion is around the earth in zonal patterns, with less than one tenth of the motion between poles and equator. The polar jet stream (part of the meridional circulation of air between poles and equator) is of importance, since it tends to meander and constantly change position. This changeable front and the resultant disturbance along the front greatly affect atmospheric circulation. The rotation of the earth determines prevailing direction of persistent winds and the prevailing direction of ocean currents, both of which contribute to weather effects.

The earth's tilt (a fixed angle of $23\frac{1}{2}°$) in relation to the plane of its path around the sun further modifies the angle at which the sun's rays strike the earth. This tilt also accounts for the four seasons and their variable weather, because entire areas of the earth are tipped toward or away from the sun for half a year at a time.

the earth's atmosphere

The atmosphere is a fluid mixture of gases surrounding the earth in several layers of varying thickness and density. The layer nearest the earth, the troposphere, extends from the earth's surface to about 10 miles or 6.25 kilometers (km). The lower troposphere (up to 2 km) is of most interest in air pollution meteorology. The atmosphere is held to earth by gravitational pull and at sea level air pressure reaches nearly 15 lb/in.^{2}, diminishing rapidly to $7\frac{1}{2}$ lb at 18,000 ft.

Temperature in the lower troposphere tends to become colder with altitude. As air is warmed by reradiation of heat from the earth's surface, molecules become agitated and push away from one another, causing the air to expand. As the hot air expands, it becomes less dense and exerts less pressure than the cold air. As the hot air expands, the cooler air next to it, under greater pressure, pushes sideways, forcing the warmer air upward. These temperature differences create imbalance in pressure, causing air to flow from areas of high pressure (near the earth) to areas of low pressure (at increased altitudes from the earth). However, because of the varying character of the underlying surface and radiation at the surface, temperature may actually increase with height, causing an *inversion*. In the upper troposphere, decrease of temperature with height averages 4 to 8°/km.

Temperature also varies horizontally, particularly with latitude. Due to the shape of the earth, warmer temperatures are found near the equator and colder temperatures are found near the poles. Within a given latitude, temperatures of course vary from hot to cold with an increase in altitude, as mentioned previously. Continents lose or gain heat rapidly, whereas the seas lose or gain heat gradually. Continents have great extremes in temperature, becoming warmer in the summer and colder in the winter, whereas the oceans maintain a more moderate temperature throughout the year. Water requires more heat per unit weight to increase its temperature 1° (specific heat) than any other common substance. This means water can absorb more heat per unit weight with less change in its own temperature. It acts as a temperature stabilizer of the environment. The land heats quickly, expanding the air and creating a low-pressure area that is filled by cooler air blowing in from the sea. When the sun sets, the land cools and the wind blows seaward.

Precipitation and humidity in the atmosphere have major effects on temperature and winds. Water vapor is always present in the atmosphere, and its effect on the balance of incoming and outgoing radiation was discussed under radiation. The *condensation* temperature of the atmosphere is below the boiling point of water, yet water is volatile enough to evaporate (change from liquid to gas) or sublimate (change from solid to gas) at atmospheric temperatures and pressures. One measure of the amount of moisture in the air is the dew-point temperature at which saturation is reached if the air is cooled at a constant atmospheric pressure without addition or loss of moisture. At the earth's surface, when night comes the ground yields up the day's heat, and with this temperature drop airborne water vapor condenses to form dew or frost, depending on the degree of cold. Closely akin to these is fog, which is condensation on invisible particles in the air. Formations of clouds, fog, hail, rain, and frost are important measurable meteorological factors that have effects on air pollution. Rain and snow carry large amounts of both particulate and gaseous pollutants out of the atmosphere and into the soils and waters of the earth. Trees and grasses act like the fibers of an enormous filter mat to collect particles and some gases. Some investigations indicate that air pollutants are effective cloud-forming agents, since their attraction for water vapor permits condensation and ice-crystal formation on dust nuclei. This in turn results in increased potential precipitation.

geophysical features of the earth's surface

Geophysical land forms (mountains, valleys, oceans, continents) over which air masses travel have a great effect on weather and meteorological conditions. As an example, a valley tends to channel wind flow along the valley axis. During the evening hours, radiation of heat from the earth's surface and consequent cooling of the ground and air adjacent to the ground cause density changes. Along the valley slopes the air becomes more dense, because it is further from the radiating surfaces than air above the valley floor. This causes the colder air to flow downward into the valley as a slope wind. As the valley air becomes colder, this cold air has a tendency to move downhill along the valley axis in a valley wind. On a clear day with light winds, the heating of the valley may cause up-slope and up-valley winds due to the tendency for heated air to rise in altitude, as previously discussed, and follow the reverse direction of evening tendencies.

Along shorelines of bodies of water, the land becomes heated on summer days with clear skies and light winds, causing a cooling sea or lake breeze to flow landward. At night, rapid radiational cooling of the land causes lower temperatures over land than over the water and creates a land breeze that moves toward the water. This was previously explained as a phenomenon related to the heat-exchange characteristics of water as compared to other substances.

When hills surrounding a coastal area form a basin, an intense pollution problem may be created. The cool air blows inland from the sea, pushing the air pollutants toward the mountains, where they become trapped as the cool air moves over the top of the polluted air to form a lid that prevents pollutant escape. When fog is formed in this process, smog conditions occur, as in the Los Angeles Basin. This smog is formed because under clear skies at night the ground loses much heat because of outgoing radiation and the air in contact with the ground is cool. When the air is sufficiently humid, the cooling will bring the air to the saturation point and fog will form. This fog combining with the smoke containing air pollutants is referred to as *smog*. The top of the layer of smog will radiate essentially as a blackbody and cool further, thus forming an inversion layer directly above the smog, trapping concentrated air pollutants below it.

stability and instability

It has been pointed out that air moves because some parts of it have more energy than others, that is, are hotter than others. The relationship of pressure to changes in temperature has also been covered. Pressure and temperature changes have a combined influence on the stability of the atmosphere. Stability is highly dependent upon the vertical distribution of temperature which will be discussed more thoroughly next.

With an increase in altitude, there is a pressure decrease that allows a parcel of air to expand and cool. If this expansion occurs without a loss or gain of heat to the parcel, the change is said to be *adiabatic.* A parcel moving downward under these conditions will encounter higher pressures, will contract, and tend to become warmer. The *rate of temperature increase or decrease with height is the dry adiabatic process lapse rate,* which is 5.4°F/1,000 ft or approximately 1°C/100 m.

The manner in which temperature changes with height at any one time is referred to as the *environmental or prevailing lapse rate.* This is principally a function of the temperature of the air and of the surface over which it is moving and the rate of exchange of heat between the two. For example, during clear days in midsummer, the ground will be rapidly heated by solar radiation, resulting in rapid heating of the layers of the atmosphere nearest the surface; but further aloft the atmosphere will remain relatively unchanged. At night, radiation from the earth's surface cools the ground and the air adjacent to it, resulting in only slight decrease of temperature with height. Because of varying character of the underlying surface and radiation at the surface, temperature may increase with height.

If the temperature decreases more rapidly with height than the dry adiabatic lapse rate, the air has a *superadiabatic or strong lapse rate* and the air is unstable. If a parcel of air is forced upward, it will cool at the adiabatic lapse rate, but, being warmer than the environmental air, it will continue to rise. Similarly, a parcel forced downward will heat at the adiabatic lapse rate, but will remain cooler than the environment and will continue to sink.

If the environmental lapse rate decreases with height at a rate less than the dry adiabatic lapse rate (*subadiabatic or weak lapse rate*), a lifted parcel will be cooler than the environment and will sink; a descending parcel will be warmer than the environment and will rise. This may also be referred to as the *common adiabatic lapse rate* (decrease in temperature versus height from 0.1 to 5.3°F/1,000 ft of rise).

When the temperature at the bottom of a layer of air is cooler than the temperature at the top or when temperature increases with altitude, we have an *inversion*. When the lapse rate is zero for a layer of air, it is termed *isothermal*; there is no temperature change with height within the layer.

These varying lapse rates based on temperature variation with height determine atmospheric stability, which in turn has a major effect on the manner in which stack effluents diffuse in the atmosphere. The relation between lapse rates and the behavior of smoke plumes is shown later in this chapter. Figure 12–1 shows the lapse rates depicted on a temperature profile.

EFFECTS OF METEOROLOGICAL PARAMETERS ON TRANSPORT AND DIFUSSION

The air pollution cycle can be considered to consist of three phases: release of air pollutants at the source, transport and diffusion in the atmosphere, and reception of air pollutants in reduced concentrations by people, plants, animals, and inanimate objects. Although meteorological factors have some influence on all three of these phases,

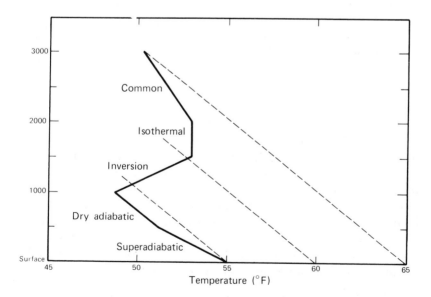

Fig. 12–1 Lapse rates on temperature Profile (Courtesy of OMD-APCO-EPA).

they are greatest during the diffusion and transport phase. The factor of most concern during transport and diffusion of pollutants is the wind effect on those pollutants.

wind

Wind is air in motion in three dimensions. Previously in this chapter the wind effects due to radiation from the sun, the earth's shape and movement, temperature and pressure variations, and geophysical features of the earth's surface have been covered. Our main concern regarding wind and its effects on distribution of air pollution in the transport and diffusion phase is in the determination of wind speed and wind direction.

Wind speed is measured in *miles per hour* (mph) or *meters per second* (m/s). The effect of wind speed is twofold. One effect is that wind speed will determine the travel time of a pollutant from source to receptor (man, plant, animal affected). If a receptor is located 1,000 m downwind from a source and the wind speed is 5 m/s, it will take 200 s for the pollutants to travel from the source to the receptor (1,000 ÷ 5 = 200 s). The other effect is the amount of the pollutant dilution in the windward direction. For example, if a continuous source is emitting a certain pollutant at the rate of 10 g/s and the wind speed is 1 m/s, a downwind plume 1 m in length will contain 10 g of pollutant, since 1 m of air moves past the source each second. The concentration of air pollutants is inversely proportional to wind speed, since a fast-moving wind quickly disperses the pollutants, and this diffusion within the air mass reduces the concentration of pollution. A general description of winds based on the Beaufort scale of wind-speed equivalents is shown in Table 12–1.

Wind direction is the direction *from which the wind blows*. If the wind direction measured is representative of the wind movement at the height at which a pollutant is released, the mean direction of the wind (midpoint of all measurements) will be indicative of the direction of travel of the pollutants (e.g., a NW wind will move pollutants to SE of the source). A *wind rose* is a graphic display of the distribution of wind direction experienced at a given location (i.e., prevailing wind direction over a period of time, usually a month). An example of a wind rose is shown in Fig. 12–2.

Wind variability includes deviations from the mean velocity of the wind, both vertical and horizontal, which cause *atmospheric tur-*

TABLE 12-1   The Beaufort Scale of Wind-Speed Equivalents

General Description	Specifications	Limits of Velocity 33 Feet (10 m) above Level Ground (Miles per Hour)
Calm	Smoke rises vertically.	Under 1
	Direction of wind shown by smoke drift but not by wind vanes.	1 to 3
Light	Wind felt on face; leaves rustle; ordinary vane moved by wind.	4 to 7
Gentle	Leaves and small twigs in constant motion; wind extends light flag.	8 to 12
Moderate	Raises dust and loose paper; small branches are moved.	13 to 18
Fresh	Small trees in leaf begin to sway; crested wavelets form on inland waters.	19 to 24
	Large branches in motion; whistling heard in telegraph wires; umbrellas used with difficulty.	25 to 31
Strong	Whole trees in motion; inconvenience felt in walking against the wind.	32 to 38
	Breaks twigs off trees; generally impedes progress.	39 to 46
Gale	Slight structural damage occurs (chimney pots and slate removed).	47 to 54
	Trees uprooted; considerable structural damage occurs.	55 to 63
Whole gale	Rarely experienced; accompanied by widespread damage.	64 to 75
Hurricane		Above 75

Extracted from Air Pollution Manual Part 1—Evaluation, American Industrial Hygiene Assoc., 14125 Prevost, Detroit, Michigan, 1960.

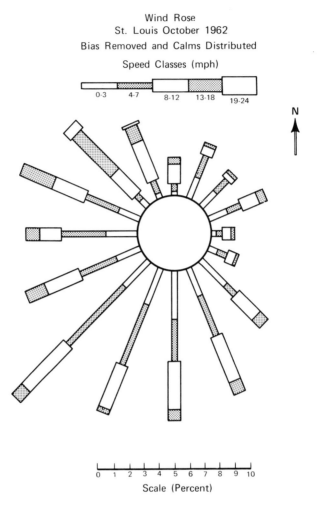

Wind Rose
St. Louis October 1962
Bias Removed and Calms Distributed

Speed Classes (mph)

0-3   4-7   8-12   13-18   19-24

N

0 1 2 3 4 5 6 7 8 9 10
Scale (Percent)

Fig. 12–2 Wind rose (Courtesy of OMD-APCO-EPA).

bulence. The roughness of the surface over which the air passes (trees, shrubs, buildings, terrain features) induces *mechanical turbulence.* The height and spacing of the elements causing the roughness will affect turbulence. In general, the higher the roughness, the greater the turbulence. In addition, mechanical turbulence increases as wind speed increases.

Varying temperatures create *thermal turbulence,* which contributes to wind variability and its effect on atmospheric stability. *Diurnal variations* (occurring every day) are another factor contributing to wind variability. During the daytime, solar heating causes turbulence and vertical motions to be at a maximum. This causes the maximum amount of momentum exchange between various levels in the atmosphere. Because of this, variation of wind speed with height is minimal during the daytime. At night, vertical motions are minimal and friction effects do not affect as deep a layer of the atmosphere as during the day.

Frontal trapping is a term used to describe the trapping of air pollution beneath the inversions that accompany frontal systems (warm or cold). Relatively high pollutant concentrations may build up, as they do in the Los Angeles area. Warm fronts are the worst, since they usually move more slowly due to pressure gradient force. Warm fronts also have a more gradual sloping frontal surface than a cold front, due to their slower speed. In addition, the low-level and surface wind speeds ahead of a warm front will usually be lower than for a cold front because of less pressure exerted behind them.

POLLUTANT CONCENTRATION VARIATION

The manner in which stack effluents diffuse is primarily a function of the stability of the atmosphere. There are three general methods that can be used to maximize the dilution capacity of the atmosphere: the area of stack design, the area of control through zoning and land usage, and the reduction of source operation to utilize minimum meteorological conditions for pollutant diffusion.

In regard to *stack design*, several factors have been determined on the basis of observation of past design that can be used as aids in future stack design or to predict pollutant concentrations to be expected from pollutant stacks. Generally speaking, under the same meteorological conditions, ground-level pollutant concentration at a given point downwind from a stack will be reduced as the stack height is increased. This is due to the fact that pollutants emitted from the stack are allowed more time to become diffused in the atmosphere before reaching the ground.

The effective stack height (H') is the physical height of the stack (H) plus the height differential due to the exit velocity (hv) of the stack gases and the buoyancy (ht) of the stack gases, as shown in Fig. 12–3.

Experience has proved that if the exit velocity is higher than the outside wind speed by a ratio of 4 to 3, the exit velocity factor (hv)

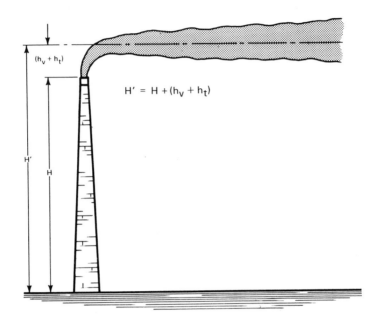

$$H' = H + (h_v + h_t)$$

Fig. 12–3 Definition of effective stack height (Courtesy of OMD-APCO-EPA).

will be positive. Stack gas temperature determines the buoyancy of the smoke plume emitted from the stack. Experience has shown that if the stack gas temperature is higher than the ambient air temperature the buoyancy factory (h_t) will be positive. When heat is absorbed by condensation of water vapor, buoyancy could be negative, a condition which could result from using a wet collector device on a stack. Downwash (a downward movement of pollutants caused by downward movement of the atmospheric air carrying the pollutant) may result from a negative buoyancy factor, which may lead to a lowered effective stack height. Downwash may also result if negative pressure is created on the lee side (side sheltered from the wind) of the stack as a result of general air movement. The stack effluent is drawn into this low-pressure area and pollutants are deposited near the stack.

Atmospheric stability is a very important factor in dilution of pollutants in the atmosphere. Obstructions in the area like buildings, trees, or hills will create mechanical disturbances that cause the wind flow to fluctuate and become rough at the lower levels. Thermal disturbance due to solar heating on a hot, sunny day with very low wind speeds can bring portions of a smoke plume from a tall stack down to the ground within very short distances. Various types of smoke plumes are depicted in Fig. 12–4 with an accompanying temperature

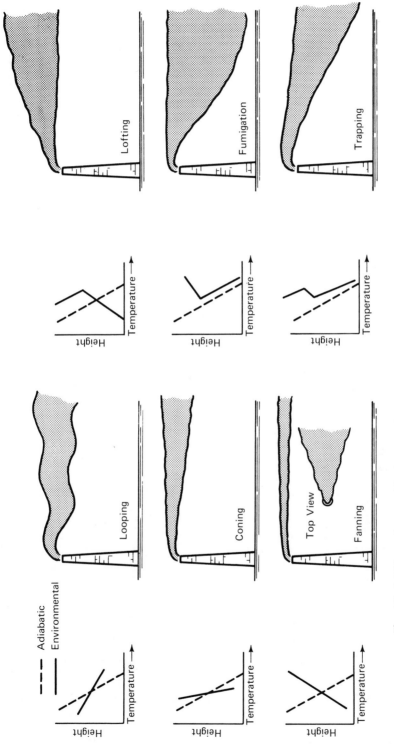

Fig. 12-4 Effect of atmospheric stability on plume behavior (Courtesy of OMD-APCO-EPA).

235

TABLE 12-2 Plume Behavior and Related Weather

Description of Visible Plume	Typical Occurrence	Temperature Profile—Stability	Associated Wind and Turbulence	Dispersion and Ground Contact
		1. Looping		
Irregular loops and waves with random sinuous movements; dissipates in patches and relatively rapidly with distance.	During daytime with clear or partly cloudy skies and intense solar heating; not favored by layer-type cloudiness, snow cover or strong winds.	Adiabatic or super-adiabatic lapse rate—unstable.	Light winds with intense thermal turbulence.	Disperses rapidly with distance; large probability of high concentrations sporadically at ground relatively close to stack.
		2. Coning		
Roughly cone-shaped with horizontal axis; dissipates farther downwind than looping plume.	During windy conditions, day or night; layer-type cloudiness favored in day; may also occur briefly in a gust during looping.	Lapse rate between dry adiabatic and isothermal—neutral or stable.	Moderate to strong winds; turbulence largely mechanical rather than thermal.	Disperses less rapidly with distance than looping plume; large probability of ground contact some distance downwind; concentration less but persisting; longer than that of looping.
		3. Fanning		
Narrow horizontal fan; little or no vertical spreading; if stack is	At night and in early morning, any season; usually associated with	Inverted or isothermal lapse rate—very stable.	Light winds; very little turbulence.	Disperses slowly; concentration aloft high at relatively great distance down-

Type	Lapse rate	Wind & turbulence	Conditions / time	Appearance	Ground contact
(continued)			inversion layer(s); favored by light winds, clear skies and snow cover.	high, resembles a meandering river, widening but not thickening as it moves along; may be seen miles downwind; if effluent is warm, plume rises lowly, then drifts horizontally.	wind; small probability of ground contact, though increase in turbulence can result in ground contact; high ground level concentrations may occur if stack is short or if plume moves to more irregular terrain.
4. Lofting	Adiabatic lapse rate at stack top and above; inverted below stack—lower layer stable, upper layer neutral or unstable.	Moderate winds and considerable turbulence aloft; very light winds and little or no turbulence in layer below.	During change from lapse to inversion condition; usually near sunset on fair days; lasts about an hour but may persist through night.	Loops or cone with well defined bottom and poorly defined, diffuse top.	Probability of ground contact small unless inversion layer is shallow and stack is short; concentration high with contact but contact usually prevented by stability of inversion layer; considered best condition for dispersion since pollutants are dispersed in upper air with small probability of ground contact.
5. Fumigation	Adiabatic or super-adiabatic lapse rate at stack top and below; isothermal or inverted lapse rate	Winds light to moderate aloft, and light below; thermal turbulence in lower layer, little turbulence in upper layer.	During change from inversion to lapse condition; usually nocturnal inversion is being broken up through warming of	Fan or cone with well defined top and ragged or diffuse bottom.	Large probability of ground contact in relatively high concentration, especially after plume has stagnated aloft.

TABLE 12-2 (Cont'd)

Description of Visible Plume	Typical Occurrence	Temperature Profile— Stability	Associated Wind and Turbulence	Dispersion and Ground Contact
	ground and surface layers by morning sun; breakup commonly begins near ground and works upward, less rapidly in winter than in summer; may also occur with sea breeze in late morning or early afternoon.	above—lower layer unstable or neutral, upper layer stable.		
		6. Trapping		
Cone with well defined top and increasing concentration downward; dissipates farther downwind than fumigation plume.	When an inversion occurs aloft such as a frontal or subsidence inversion, a plume released beneath the inversion will be trapped beneath it.	Inverted lapse rate, very stable.	Winds light to moderate aloft and light below; moderate turbulence in lower layer.	Larger probability of ground contact in relatively high concentration.

Extracted from Air Pollution Manual Part 1—Evaluation, American Industrial Hygiene Assoc., 14125 Prevost, Detroit, Michigan, 1960.

profile that corresponds to conditions normally present when these plumes occur. More detailed plume behavior and related weather are listed in Table 12–2. The relationship between lapse rates, stability, and plumes is depicted in Fig. 12–6. From these figures it can be seen that with a very stable atmosphere the gas effluent from the stack should form a *fanning* plume. In the morning with the warming of the ground surface, the atmospheric stability becomes unstable, and as this unstable layer of air reaches the fanning plume, the highly concentrated pollutants are forced to the ground by the turbulence, creating a *fumigating* plume when winds are light to moderate aloft.

When an inversion occurs aloft in an otherwise stable atmosphere with light winds aloft and moderate turbulence below, a *trapping* plume may be found. A clear hot day accompanied by light winds may produce a *looping* plume. With moderate to strong winds and turbulence largely mechanical rather than thermal, a *coning* plume may be expected. Toward sunset on a fair day with considerable turbulence aloft and little turbulence below, a plume may become *lofting* as adiabatic lapse rate changes to inversion conditions.

It is useful to know the height to which vertical mixing of pollutants in the atmosphere is likely to occur. The maximum mixing depth (MMD) can be estimated from a temperature profile by plotting the maximum surface temperature for the day on the morning profile (e.g., 61°F). From this point, draw a line parallel to a dry adiabating line to the point where it intersects the early morning profile. Read the height above ground at this point. This is the MMD for the day. Figure 12–5 shows an example with MMD of approximately 1,600 ft.

A summary of the relationships between the type of lapse rates, plume type, and type stability is depicted in Fig. 12–6.

METEOROLOGICAL INSTRUMENTS

Meteorological instruments for measurement of atmospheric variables that affect the diffusion and transport of air pollutants are required in air pollution programs. Sometimes meteorological data can be obtained from weather bureau offices, airport stations, or other sources; however, existing instruments may not provide enough detail, may not be representative of the area in question, or may not measure variables such as turbulence. In this case, it may be imperative to operate a meteorological station as part of the functions of an air pollution control agency to ensure the collection of all required data at areas that are representative of the area in question.

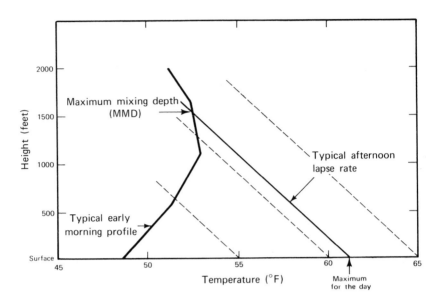

Fig. 12–5 Maximum mixing depth (Courtesy of OMD-APCO-EPA).

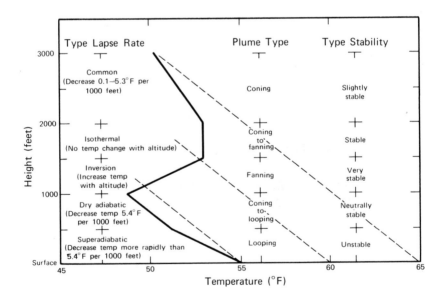

Fig. 12–6 Summation of lapse rates, stability, and plumes (Courtesy of OMD-APCO-EPA).

As an index to turbulence, the three most important variables are *wind speed, wind direction,* and *atmospheric stability.* Of secondary importance is the measurement of *humidity, temperature, precipitation, solar radiation,* and sometimes *visibility* and *illumination.*

Wind speed is usually measured by a rotation anemometer—a propeller or windmill mounted on one end of a horizontal shaft oriented into the wind by a wind vane on the opposite end of the shaft, or three or four cups rotating freely around a vertical shaft. The wind speed may be transmitted to a recorder by mechanical, optical, electrical, or magnetic means. Figure 12–7 shows a wind-speed instrument (three-cupped anemometer).

Wind direction, which is the direction from which the wind is blowing, may be measured by three types of wind vanes: flat plated and dynamically shaped or splayed vanes. The flat plate is adequate and costs less to purchase. Wind direction may be recorded by mechanical linkage or by electrical circuits. Recorders are available to register both wind speed and 16 to 20 points of wind direction, which is considered sufficiently accurate for all wind-direction readings.

Fixed-position wind systems should be on a mast or pole 10 m (33 ft) above the ground. Wind speed and direction are affected by buildings a distance equal to the height of the building upwind, a

Fig. 12–7 Climet wind vane and anemometer (Courtesy of Climet Instrument Co.).

distance equal to the height of the building above the building, and five to ten times the height of the building on the downwind side. A rule of thumb is to place wind instruments at a distance ten times the height of the obstruction. When mounted on a building, wind instruments should be at the highest point with no obstructions; furthermore, there should be no taller buildings in the vicinity. When mounted on a tower, the boom should extend out at least one tower width. Figure 12–7 shows a wind-direction instrument on the end opposite from the anemometer.

Airborne sensors are used to average wind velocity through a given depth of the atmosphere at a particular time. A gas-filled free balloon may be used to carry a radio transmitter that is visually tracked by radio direction-finding apparatus or radar. The radiosonde, which utilizes radio soundings, is used to transmit pressure, temperature, and humidity information to the ground.

Atmospheric stability is usually determined by measuring *temperature* at two heights, subtracting the temperature of the higher height from that of the lower height, and comparing the difference to the dry adiabatic lapse rate for the height interval. Turbulence may be estimated from studying the correlation between fluctuations of wind speed and wind direction. A two-vane anemometer may be used to record both horizontal and vertical wind movements.

Temperatures at different heights may be determined from aircraft-borne sensors, captive balloons, radiosondes in free balloons, or resistance thermometers located at varying heights on radio towers. Maximum–minimum thermometers may be used to estimate maximum mixing depth for the day. A thermocouple, which measures relationships between the temperature and the electrical condition in a metal or in contacting metals, or a thermometer, which measures temperature based on the rise and fall of mercury in a graduation glass tube, is placed in an instrument shelter to record varying temperatures at ground level. Figure 12–8 shows thermometers mounted in an instrument shelter.

A *hygrothermograph* (Fig. 12–9) may be used to make a continuous recording of both temperature and relative humidity. The humidity sensor generally used is human hair, which lengthens as relative humidity increases and shortens with humidity decreases. Temperature measurements are usually made with a Bourdon tube, which is a curved metal tube containing an organic liquid. The system changes curvature with temperature, activating the pen arm for recording on graph paper. A *psychrometer* (Fig. 12–10) may be useful for spot check-

Fig. 12–8

ing instantaneous *humidity* values. Humidity measurement by a psychrometer depends upon obtaining a dry-bulb temperature and a wet-bulb temperature from a matched thermometer whose bulb is covered with a muslin wick moistened with distilled water. There must be enough air motion to cause cooling of the wet bulb due to evaporation of the water on the wick. Relative humidity is determined from dry–wet bulb readings through use of tables designed for this purpose.

Since large particles and water-soluble gases may be removed from the atmosphere by falling precipitation, precipitation measurements may be needed. A *rain gauge* may be used to record *precipitation* in the analysis of pollutant samples collected. Chemical and radioactive constituents in rainwater may be an important factor. The occurrence or nonoccurrence aspect may be more important than the quantitative amount of rainfall. Moisture in the form of dew or rain is also important in the study of air pollution effects on materials. Dew-recording devices are available. Figure 12–11 shows a rain gauge.

Solar radiation may be measured for research purposes to determine the influence of the sun's rays upon turbulence and certain

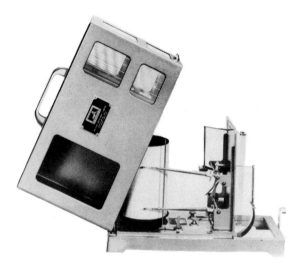

Fig. 12–9 Hygrothermograph (Courtesy of Belfort Instrument Co.).

photochemical reactions. A *pyranometer* measures total radiation, a *pyrheliometer* measures direct solar radiation, and a *sun photometer* is designed to make spot readings of radiation.

An *illuminometer* is an instrument used to measure the radia-

Fig. 12–10 Psychrometer (Courtesy of Bendix Co.).

Fig. 12–11 Clear-Vu rain gauge (Courtesy of Science Associates Inc.).

tion received on a horizontal surface at the wavelengths sensed by the human eye. A *transmissometer* is used to measure *visibility* in one direction. Usually visibility is recorded by trained observers, as described in succeeding paragraphs.

APPLICATION OF METEOROLOGICAL DATA

Meteorological data make it possible to forecast potential air pollution episodes so that emergency controls may be applied to prevent their occurrence, and help in designing stacks and locating stacks and plants where they may contribute less air pollution. The use of atmospheric trace elements has helped to study air patterns used to establish air pollution programs and to prove transport of pollutants in enforcement backup. Dispersion models (a mathematical description of the meteorological transport and dispersion processes combined with meteorological parameters during a particular period and emission inventory of a pollutant of interest) are used in urban-area air pollution control planning and improvement. Local factors pertaining to the weather can be used to assist in revising emission standards for application to a specific area.

EFFECTS OF AIR POLLUTANTS ON ATMOSPHERIC VISIBILITY

visibility

Deterioration of *visibility* is probably the first indication of air pollution of which a citizen becomes aware.

The *Glossary of Meteorology* describes visibility as "the greatest distance in a given direction at which it is just possible to see and identify with the unaided eye (a) in the daytime, a prominent dark object against the sky at the horizon and (b) at night, a known, preferably unfocused, moderately intense light source." These observations are taken primarily for aircraft operations.

Light scattering of sunlight by particulates suspended in the atmosphere (particularly when the relative humidity is below 70 percent) is the main cause of reduction in atmospheric visibility. This condition is comparable to a dirty automobile windshield. At night with no oncoming vehicle, it is not too bothersome; however, when strong light from the sun or vehicular lights strikes the dirty windshield, the driver's visibility is impaired. A concentration of 150 μg of 1-micron-diameter particles/m³ of air will reduce visibility to about 2 miles. The obscuring ability of the airborne particulates is dependent primarily on the number of particles and their size. Particles of 1 micron diameter or smaller scatter more light than larger particles, since scattering area per gram of material is greater with smaller particles.

Since light scattering is the main cause of reduction in visibility, this aspect of air pollution is manifested primarily during bright days when there is strong light to be scattered by the suspended particulates. Although the same particulate concentration may exist on a cloudy day or at night, the atmosphere may be referred to as "clear" if no haze is evident to the naked eye.

pollutants contributing to impairment of visibility

Major contributors to impairment of atmospheric visibility are hygroscopic particulates (particulates that collect water from the atmosphere and form a mist). Some hygroscopic particles, such as ether—soluble organics and sulfur trioxide from industry, are produced by man; others come from natural sources such as sea salt spray.

Opaque agglomerates such as carbon, tar, and metal particles in the atmosphere also contribute to reduced visibility. Crystalline compounds such as iron, aluminum, silicon, and calcium, which may exist as sulfates, nitrates, chlorides, and fluorides, also contribute to light scattering.

METHODS OF MEASUREMENT OF VISIBILITY

In air pollution studies, U.S. Weather Bureau visibility observations are frequently used. Several prominent dark objects against the sky at the horizon are selected and their known distance plotted. During daylight hours the observer determines the greatest distance in a given direction at which time it is barely possible to see and identify the selected objects with the unaided eye. The observer then resolves these distances into a single value, which is lowest visibility over the half of the horizon (not necessarily continuous) with the greatest visibility. He records this as the prevailing visibility. At night, a known moderately intense light source (preferably unfocused) is used to determine visibility.

As an example, the observer may divide the horizon into quadrants around his location. Visibility NE is read as 7 miles, SE as 10 miles, SW as 5 miles and NW as 5 miles. Visibility would be reported as 7 miles, the average of the four quadrants expressed in the nearest whole number. When prevailing visibility is less than 7 miles, the observer must also record the cause of restriction, such as haze, smoke, rain, or the like. Precipitation is always reported whether or not visibility is affected.

To estimate roughly the reduction of visibility due to scattering of light by particulate matter, attenuation of light may be considered a function of particle density, radius, concentration, and light-scattering efficiency. A formula sometimes used to determine visibility is

$$V = \frac{2.5 \times P \times a}{K \times M}$$

where

V = visibility, miles

P = density particles, g/m^3

a = radius of the particles, microns

K = scattering coefficient or efficiency of the particles—a function of the ratio of particle size to wavelength

M = concentration of particles, mg/m³

Example:

$$P = 1.5 \text{ g/m}^3$$

$$a = 0.5 \text{ } \mu$$

$$M = 0.6 \text{ mg/m}^3$$

$$K = 3.6$$

$$V = \frac{2.5 \times 1.5 \times 0.5}{3.6 \times 0.6} = \frac{7}{8} \text{ mile}$$

K is determined from Fig. 12–12, which has been constructed with an assumed index of fraction $m = 1.5$ for haze. Note particle size 0.25 to 0.6 μ has the highest scattering efficiency; this size range is very close to that of visible-light wavelengths, 0.4 to 0.7 μ.

Using the formula and interpreting K from Fig. 12–12, visibility observations of less than 1 mile could be anticipated when heavy suspended particulate loadings occur. Adverse levels of particulates are considered to have been reached when visibility is reduced to less than 3 miles when relative humidity is less than 70 percent.

ECONOMIC SIGNIFICANCE OF VISIBILITY IMPAIRMENT

Reduction in visibility due to air pollution creates an economic burden because of increased requirements for electricity in homes, in business establishments, and in the streets when sunlight cannot penetrate the haze. Airport operations are slowed down because of delays in air traffic, which add to operational costs, inconvenience passengers, and pose additional hazards to safety that may result in death, personal injury, or property damage.

Highway traffic is impaired when the motorist's vision is limited. Traffic arteries may be clogged to a standstill; accidents causing bodily injury, death, and increased property damage may occur. If the community is near a harbor, ship traffic may suffer in the same way as air and highway traffic. Serious indirect costs may arise if airports have to

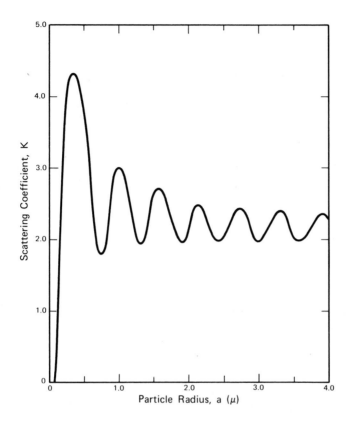

Fig. 12–12 Variation of scattering coefficient with particle size for visible light ($\lambda = 5500$ Å). (Courtesy of OMD-APCO-EPA).

be relocated or enlarged to handle slower traffic. Additional highway requirements and increased insurance rates and other indirect economic costs may result from reduction in visibility if it becomes exceedingly severe or frequently occurring.

REFERENCES

Air Pollution Manual, Part 1—Evaluation, American Industrial Hygiene Association, 14125 Prevost, Detroit, Michigan, (1960).

Bulrich, D., Scattered Radiation in the Atmosphere and the Natural Aerosol, *Advances in Geophysics*, Vol. 10, ed. by H. E. Landsberg and J. Van Mieghem. New York: Academic Press, Inc., 1964.

Huschke, R. E., ed., Glossary of Meteorology American Meteorological Society, Boston, 1959.

Magill, P. L., F. R. Holden, and Charles Ackley, eds., *Air Pollution Handbook*. New York: McGraw-Hill Book Company, 1956.

Nadler, A. C., *The Atmosphere*. New York: Scientists Institute for Public Information, 1970.

Stern, A. C., *Air Pollution*, Vol. 1. New York: Academic Press, Inc., 1962.

U.S. Department of Health, Education, and Welfare, *Workbook of Atmospheric Dispersion Estimates*, PHS (NAPCA), Pub. No. 999-AP-26. Washington, D.C.: U.S. Government Printing Office, 1969.

RECOMMENDED FILMS

TF-110   Principles of Global Weather Circulation

MA-9, 10, 11, 12, 13   Air Pollution Meteorology Equipment

Film slides—Meteorological Instrumentation
    Available: Distribution Branch
                 National Audio-Visual Center (GSA)
                 Washington, D.C. 20409

QUESTIONS

1/ What term is used by meteorologists to describe the temperature change in the atmosphere that occurs with increasing height?

2/ What is the lapse rate that is the dividing line between stable and unstable atmospheric conditions?

3/ What type plume from an elevated source contains the highest concentration of pollutant?

4/ What type graphical display is used to estimate the stability of the atmosphere?

5/ What is meant by the "greenhouse" effect?

6/ Describe an inversion and how it may effect air pollution.

7/ What are the three phases of an air pollution cycle? Which phase is influenced most by meteorological data?

8/ What are the three types of turbulence that affect wind variability?

9/ What are the three general methods that can be used to maximize the dilution capacity of the atmosphere?

10/ What are the three most important meteorological variables to be measured? What instruments may be used to measure these variables?

11/ What are four other variables of less importance?

12/ What is the main cause of lack of visibility due to air pollution?

13/ When is this lack of visibility most apparent, day or night?

14/ What are some types of pollutants of major concern in reduction of visibility?

15/ How does the U.S. Weather Bureau determine visibility at night and during the daytime?

16/ If visibility is read as NE quadrant 7 miles, SE quadrant 6 miles, SW quadrant 4 miles, NW quadrant 8 miles, what prevailing visibility is reported?

17/ If density of atmospheric particles is determined as 210 g/m^3, radius of particles 1.0 μ and concentration of particles as 0.5 mg/m^3, what is the rough estimate of visibility?

18/ What are some of the economic aspects of reduced visibility produced by air pollution?

PROBLEMS

1/ With a receptor 500 m downwind from a source and a wind speed of 10 m/s, how long will it take for the pollutant to reach the receptor? If the pollutant is being emitted at the rate of 5 g/s, what will a downwind plume 1 m in length contain in pollutant weight?

2/ Obtain data from a weather bureau pseudoadiabatic chart and determine maximum mixing height during a given day.

3/ Obtain wind data from a U.S. Weather Bureau station and plot a wind rose.

4/ Set up a weather station and learn to read the various meteorological instruments.

Student Objectives

—*To understand the basic objectives of an air resources management program and the elements of such a program.*
—*To understand the federal role and the state role in establishing air resources management programs.*
—*To learn the sequence and relationship between air quality criteria, air quality standards, and air pollution regulations.*
—*To recognize the difference between ambient air quality standards and emission standards.*

Air resources management (or *air quality management*) is the effort to abate or reduce existing air pollution and to prevent future pollution. Since the aim of science is to seek objectivity and truth and to pursue knowledge based on concrete evidence and fact, air resources management (ARM) is a science. It is not, however, a pure science free from emotion, intuition, and philosophy. Because of an ever-expanding population, which is seeking a more favorable environment, human emotions, and the philosophy of human nature, the good life and man's place in the scheme of things cannot be overlooked when seeking solutions to air pollution problems. Therefore, *ARM may be considered as an applied science* that takes advantage of all available technological problem-solving approaches to fight and prevent air pollution.

It was pointed out in Chapter 2 that to solve the air pollution problem and thereby ascertain the parameters of an ARM program, "good" air must be defined. Then the type, emission source, and ill effects of "bad" air need to be determined. In the search for an answer to the question, "What quality air is most desirable?," the federal government has consolidated all available scientific knowledge related to air pollution into *descriptive* air quality criteria pamphlets. These criteria pamphlets are used to develop *air quality standards* that *prescribe* the quality of air desired by federal or state agencies. Finally, *air pollution control regulations* must be enacted to provide a legal basis for attaining good air.

FEDERAL GOVERNMENT AIR POLLUTION CONTROL LEGISLATION

The first air pollution control act passed in 1955 provided limited guidance to the states regarding air pollution problems. In 1963, the Clean Air Act was passed, followed by the Air Quality Act of 1967 (Public Law 90–148). These acts authorized extended research into the nature and extent of the nation's air pollution problem, authorized grants to assist states in setting up air pollution control programs, established fines for removal of auto antipollution devices, and placed primary responsibility for prevention and control of air pollution sources with state governments with a provision that the federal government could step in if states failed to act. Initially, eight broad atmospheric areas of the nation were defined in terms of the influence of climate, meteorology, and topography on the capacity of the air to dilute and dispense pollution. A *national air-sampling network* was set up to monitor the air pollution in each of these areas.

Under the Clean Air Amendments of 1970 (Public Law 91–604) the following major provisions were added:

air quality control regions

The Environmental Protection Agency (EPA) is responsible for designating *air quality control regions* based on climate, meteorology, topography, urbanization, and other factors affecting air quality conditions in each area. As of January 1972, the EPA had designated 250 of these regions, some covering only part of a state and some covering portions of several states.

To decentralize federal assistance related to air pollution control matters, the EPA has also established 10 *regional air pollution control directors*. Their offices and areas of responsibility are listed in Table 13–1. (These regional EPA offices should not be confused with the quality control regions mentioned in the previous paragraph.)

air quality criteria documents

The EPA is required by law to develop air quality criteria that reflect the adverse effects of air pollution on man and his environment. The criteria are *descriptive* because they specify a certain pollutant level and the harmful effects that result if this level is exceeded. Air quality criteria must take into consideration

1/ Chemical and physical characteristics of pollutants.

2/ Techniques for measuring pollutant characteristics.

3/ Time of exposure to a specific pollutant.

4/ Meteorological effects that influence dilution and dispersion.

5/ Susceptibility of receptors to ill effects.

6/ Responses of receptors to ill effects.

Criteria documents are expected to cover 50 or more pollutants. Paralleling these criteria documents are companion volumes on control techniques for each of the pollutants covered in criteria documents. These control techniques were covered in Chapters 11 and 12. Air quality standards develop from air quality criteria.

national ambient air quality standards

Atmospheric (ambient) air quality standards are *prescriptive*; they prescribe pollutant exposures for those contaminants covered by air quality criteria and control technique pamphlets. Air quality

TABLE 13-1   Regional Air Pollution Control Offices

Region 1	Boston, Massachusetts	(Connecticut, Maine, Massachusetts, New Hampshire, Rhode Island, Vermont)
Region 2	New York, New York	(New Jersey, New York, Puerto Rico, Virgin Islands)
Region 3	Philadelphia, Pennsylvania	(Delaware, District of Columbia, Maryland, Pennsylvania, Virginia, West Virginia)
Region 4	Atlanta, Georgia	(Georgia, Alabama, Florida, Mississippi, Kentucky, North Carolina, South Carolina, Tennessee)
Region 5	Chicago, Illinois	(Illinois, Indiana, Minnesota, Michigan, Ohio, Wisconsin)
Region 6	Dallas, Texas	(Texas, Arkansas, Louisiana, New Mexico, Oklahoma)
Region 7	Kansas City, Missouri	(Missouri, Iowa, Kansas, Nebraska)
Region 8	Denver, Colorado	(Colorado, Montana, North Dakota, South Dakota, Utah, Wyoming)
Region 9	San Francisco, California	(Arizona, California, Hawaii, Nevada, Guam, American Samoa)
Region 10	Seattle, Washington	(Alaska, Idaho, Oregon, Washington)

standards define a *desired* limit on specific pollutant levels in the air. This goal includes long-term values for exposure (e.g., annual average concentrations as well as highest 24-h concentrations in a year) and one or more short-term values (e.g., highest 1-h average concentration allowable during a year). The pollutant measurement technique and analytical method must be specified for each standard. Averaging times are expressed in single measurements (1–h, 8–h, 24–h). Pollutant concentrations are expressed in micrograms per cubic meter ($\mu g/m^3$) or by an older expression of parts per million (ppm).

Air quality standards reflect the relationship of air pollution to human health. There are two kinds of standards—*primary* standards sufficiently stringent to protect the public health, and *secondary* standards to protect the public welfare. In this relationship, each receptor and each pollutant must be considered in terms of its concentration threshold. A threshold is the pollutant concentration below the point at which none of the receptors experience an ill effect. These effects can include acute sickness or death, insidious or chronic disease, alterations of important physiological functions, and untoward symptoms and discomfort leading to a change in residence or place of employment. The first national ambient air quality standards covered six pollutants (particulates, sulfur oxides, hydrocarbons, carbon monoxide, photochemical oxidants, and nitrogen oxides). These standards, listed in Table 13-2, were established in the April 30, 1971, Federal Register.

standards of performance

Standards of performance (or emission standards) are distinct from ambient air quality standards. These standards of performance constitute the maximum permissible emission levels for given pollutants at their source. There will be two kinds of standards: *national* standards applied to new and modified sources within categories designated by the EPA, and *state* standards that apply to existing sources within those same categories of activity. The initial list of *national* standards for major stationary sources was published in the March 31, 1971, Federal Register, covering contact sulfuric acid plants, fossil-fuel-fired steam generator plants with heat inputs of more than 250 million Btu/h, incinerators of more than 2,000 lb/h charging rate, nitric acid plants and Portland cement plants. Others will probably be included in future lists.

hazardous air pollutants

Some pollutants are more toxic than others and the EPA has been directed to set an emission standard that provides an *ample* margin of safety to protect the public health from such hazardous pollutants. Asbestos, beryllium, and mercury were the first hazardous

TABLE 13-2 National Ambient Air Quality Standards
(Based on Public Law 91-604, Clean Air Act, as amended December 31, 1970)

Pollutant	Air Quality Standard ($\mu g/m^3$) Primary	Secondary	Averaging Times	Method of Measurement	AQ Criteria upon Which Standard Is Based (NAPCA or EPA Pub.)
Sulfur oxides (measured as sulfur dioxide)	80 (0.03 ppm) 365 (0.14 ppm)	60 (0.02 ppm) 260 (0.1 ppm) 1300 (0.5 ppm)	Annual arithmetic mean Max. 24-hr. concentration not to be exceeded more than once per year Max. 3-hr. concentration not to be exceeded more than once per year	Based on Fed. Reg. Vol. 36, No. 84, 1971 Pararosaniline-gas bubbler, one 24-hr. sample every 6 days or continuous monitoring or flame photometric (App. A to Fed. Reg.)	No. AP-50
Particulates	75 260	60 150	Annual geometric mean Max. 24-hr. concentration not to be exceeded more than once per year.	High-volume sampler, one 24-hr. sample every 6 days or continuous tape sampler (App. B to Fed. Reg.)	No. AP-49
Carbon monoxide (note: in mg/m^3 *not* in $\mu g/m^3$)	10 (9 ppm) 40 (35 ppm)	10 40	Max. 8-hr. concentration not to be exceeded more than once per year Max 1-hr. concentration not to be exceeded more than once per year	Nondispersive infrared-continuous monitoring or gas chromatograph or mercury replacement (App. C to Fed. Reg.)	No. AP-62

Pollutant				Method	Reference
Photochemical oxidants	160 (0.08 ppm)	160	Max. 1-hr. concentration not to be exceeded more than once per year	Neutral potassium iodide colorimetric (gas bubbler) continuous monitoring or 1 hr. samples taken between 11 A.M. and 3 P.M. local time, 5 days per week or Chemiluminescence and Coulometric (App. D to Fed. Reg.)	No. AP-63
Hydrocarbons	160 (0.24 ppm) (For use as a guide)	160	Max. 3-hr. concentration (6 to 9 A.M.) not to be exceeded more than once per year	Flame ionization detector, chromatograph (App. E. to Fed. Reg.)	No. AP-64
Nitrogen dioxide	100 (0.05 ppm)	100	Annual arithmetic mean	24-hr. gas bubbler (Jacob-Hochheiser) 24-hr. sample once every 3 days or continuous Saltsman and coulometric (App. F to Fed. Reg.)	EPA Pub. No. AP-84

All measurements are corrected to a reference temperature of $25^\circ C$ and to a reference pressure of 760 mm of mercury (1,103.2 millibars).

Secondary standards expressed in 24-hr. concentration are established as a guide. For purposes of plan development and evaluation, a classification system to categorize regions has been established. Ambient concentration limits so defined vary somewhat from the limits shown above. Each region will eventually be so classified. At that time, above data can be adjusted for a particular region of interest.

pollutants designated with proposed standards published in the December 7, 1971, Federal Register.

auto emission controls

The act has set deadlines for controlling major emissions from motor vehicles as well as regulations pertaining to fuels and fuel additives. Chapter 11 previously covered these controls.

citizens' suits

Any citizen may bring suit against any person or corporation alleged to be violating an emission standard or other limitation applicable under the Act. Citizens may also sue the EPA for failure to perform an action required by the Act.

monitoring and public information rights

The EPA may require states and individual sources to monitor pollutant emissions, to keep records, and to submit periodic reports. All such records and reports are to be considered public information except for trade secrets.

federal enforcement

Once standards and implementation plans are in effect, the EPA is required to oversee state enforcement. Where widespread violations indicate that the state is failing to enforce a plan, the EPA may step in and enforce it.

implementation plans

Within 9 months after the EPA issues national air quality standards, each state must formulate an "implementation plan" to meet, maintain, and enforce these standards in each air quality control region within its jurisdiction. The states must hold public hearings on these plans, adopt them, and submit them to the EPA for approval.

STATE ROLE IN ESTABLISHING AIR RESOURCES
MANAGEMENT PROGRAM

Three years after implementation plans are approved by the EPA, the national *primary* standards are to be in effect in each state. Every 3 months of this 3-year period states are required to forward a progress report with data on pollutant concentrations, sampler types, sampling period, time intervals and frequency of sampling, methods of collection and analysis, sampler locations, and elevation and height of sampling stations. Every 6 months of this 3-year period air quality progress reports are to list emission limitations (control strategy), sources and surveillance systems, emergency action procedures, and intergovernmental cooperation provisions.

States must carry out approved implementation plans for limiting the emission of pollutants so as to achieve the primary standards by mid-1975. If any state fails to develop or carry out such plans, the EPA is authorized to do so.

air pollution control regulations

As part of a state air pollution control program, states must set up specific *air pollution control regulations* to provide a legal basis for enforcement of those restrictions deemed necessary to prevent, abate, and control air pollution. *A control regulation is a rule intended to limit the discharge of pollutants into the atmosphere.* State air pollution control regulations encompass the following:

1/ Definitions and air quality standards.

2/ Permits and registration procedures.

3/ Monitoring and record keeping.

4/ Sampling and testing methods.

5/ Equipment malfunctions and necessary repair.

6/ Compliance schedules and regulation provisions.

7/ Emission control standards affecting fixed sources related to

    a/ Open burning and incineration.

    b/ Visible emissions.

    c/ Odorous emissions.

    d/ Sulfur compound emissions.

e/ Nitrogen oxide.

f/ Hydrocarbon.

g/ Carbon monoxide.

8/ Emission control standards affecting mobile sources.

Regulations may be based on effects, analysis of the source, or back calculations from air concentrations to allowable emissions.

ELEMENTS OF AN AIR RESOURCES MANAGEMENT PROGRAM

The basic objective of an *air resources management program* (ARMP) should be the administration of a complete program that includes identification and analysis of specific air pollution problems of a given area, determination of desired air quality standards (although much of this has been established by the federal government), and the promulgation of rules and regulations to ensure attainment of the air quality standards established.

A complete program should be designed to achieve a coordinated effort of pollution abatement and control of both water and air resources. The standards of purity should be designed to protect human health, to prevent injury to plant and animal life, to prevent damage to public and private property, to ensure continued use of natural resources for recreational purposes, and to provide opportunities for healthy industrial and agricultural development.

identification and analysis of specific air pollution problems

The air quality criteria publications of the federal government have defined many of the air pollution problems. However, each regional, state, or county ARMP should include an *air quality monitoring* system to gather data within its jurisdictional area on a continuing basis. This system should monitor short-term variations and long-term trends for evaluation of the air pollution effects to determine which effects to prevent.

Some national air quality standards or goals have been established by the federal agency (EPA); however, continuous monitoring is required to determine if these standards are being attained. Agencies below the federal level may establish more stringent standards than the national standards and may establish standards for pollutants not included under the national standards.

Data collected by the sampling-station network of the monitoring system are used as a basis for calculating the overall source reduction needed for an area to achieve a selected *air quality standard.* The agency must have laboratories in which to analyze the data collected by its sampling stations. Sufficient funds and available qualified manpower are essential to set up, operate, and maintain the monitoring network and supporting laboratories.

Concurrent weather data gathered by meteorological stations and public surveys and polls on citizen reactions to pollution problems provide additional information. The data collected from these sources and from the sampling stations must be reduced to meaningful information that can be translated into legal as well as lay language to form the basis for *air pollution control regulations.*

source-emission inventory

This inventory must be accurate, complete, and current. All sources must be registered in order to identify potential violations and to provide planning and regulatory data. Registration can be accomplished by a survey technique involving questionnaires addressed to plant operators, by the inventory technique of direct inspection by field personnel, or by licensing, which involves both on-site inspection and compulsory submission of operational data. Under the licensing system, a plant must be registered in order to operate within the agency area. The amount of emissions from each source in the area must be determined in order to decide how much of each type contaminant each source can be permitted to emit and still achieve the desired air quality. Diffusion models and other mathematical models may be used to accomplish this.

Emission standards (standard of performance) are set at levels needed to meet *ambient air quality standards,* but must also be within the feasible range of present control technology. More stringent standards may be projected into the future.

Under circumstances where compliance with air quality standards will require time to accomplish necessary corrective measures, an agency may approve an application for a *temporary permit* to continue discharging air contaminants into the atmosphere. Such permits are usually granted only upon application from the firm responsible for the pollution and after detailed research into the validity of the request. The permit should be temporary and have an established termination date at a time when corrective action can be reasonably accomplished.

legal aspects of air pollution control

The right of society to prohibit air pollution and to regulate the sources of air pollution is firmly established in legislative and judicial precedent and in the legal power delegated to the states by the federal constitution. The statutory control of air pollution is not limited to the *correction* of existing conditions, but can be applied in a flexible and farsighted manner to *prevent* pollution hazards, disasters, or worsening pollution trends. Air pollution law is designed to protect the health and welfare of the public.

Although violations of air pollution law may result in criminal action, the basic intent of the law is to achieve standards of compliance on the part of the sources of air pollution, not to punish for its own sake. Actual control is achieved by cooperation on the part of the community, industry, and individuals.

Legal aspects of air pollution control include *apprehension, trial,* and *penalty.* These legal aspects may be based on *common law,* which is general in nature, for abating nonroutine or nuisance incidents. Common law may also be used as a basis for establishing *statute laws* at the federal and state levels or ordinances at the city or county levels. A statute law, which is more specific than an ordinance, becomes a mechanism for abatement and control of air pollution.

The method of penalty may be based on *criminal code* or *civil injunction* or a combination of the two. Whereas the penalty under criminal code is usually fine or imprisonment, penalty under a civil injunction is usually enforcement of a specific code of conduct from a source owner. In most air pollution offenses, criminal intent is not needed; proof of violation is sufficient.

The Clean Air Act provides legal status for the federal government's role in air pollution matters to include establishment of air quality criteria and control techniques, national ambient air quality standards, national emission standards, and the federal enforcement role. Provision is made to delegate much of this authority to a state when the state prepares an adequate plan to implement and enforce air quality and emission standards at least as stringent as the national standards. Certain powers are specifically retained by the federal government (e.g., jurisdiction pertaining to emission standards for new automobiles). In most cases states delegate certain powers to county or city air pollution control agencies.

air pollution control regulations

As stated previously, an air pollution regulation is a rule intended to limit the discharge of pollutants to the atmosphere and thereby achieve a desired degree of ambient air quality. In setting the regulations, one is concerned with information that can be used to justify or establish a specific limitation. In writing the regulation we are concerned with ease of enforcement, prevention of circumvention, and terminology and specifications suited to the emitter.

The statutes, ordinances, rules, and regulations for controlling and preventing air pollution generally fall into the following categories: public nuisance, maximum permissible emission standards, regulation of use or design of equipment, regulation of fuels or fuel consumption.

Public nuisance   This is a version of common law nuisance included in a law which directly makes illegal any quantity of air contaminants that has a detrimental effect on the health, comfort, and property of any considerable number of people. Because the nuisance does not incorporate specific standards, it must be established in each case on its own merits. The basis may be an immediate sensory effect, long-term effects, or ambient air concentrations where an allowable *concentration* is used as a control regulation. Examples are laws prohibiting open burning, laws controlling or prohibiting odorous emissions, or laws pertaining to allowable concentrations of sulfur dioxide or particulates in the atmosphere. Detection in each of these cases may be achieved by sampling at fixed locations in areas beyond the premises in which a source is located.

Examples of state air pollution regulations based on public nuisance version of common law are:

REGULATION NO. 1—CONTROL AND PROHIBITION OF OPEN BURNING

1.0   purpose

This regulation is for the purpose of preventing, abating, and controlling air pollution resulting from air contaminants released in the open burning of refuse or other combustible materials.

1.1   scope

This regulation shall apply to all operations involving open burning except those specifically exempted by Section 1.3.

1.2

No person shall cause, suffer, allow, or permit open burning of refuse or other combustible material except as may be allowed in compliance with Section 1.3, or except those covered by a permit issued by the Board under Section 143–251.1 (c) of the Act or the regulations of a duly certified local air pollution control program having jurisdiction.

1.3   permissible open burning

While recognizing that open burning contributes to air pollution, the Board is aware that certain types of open burning may reasonably be allowed in the public interest; therefore, the following types of open burning are permissible as specified if burning is not prohibited by ordinances and regulations of governmental entities having jurisdiction. The authority to conduct open burning under the provisions of this section does not exempt or excuse any person from the consequences, damages, or injuries which may result from such conduct nor does it excuse or exempt any person from complying with all applicable laws, ordinances, regulations, and orders of the governmental entities having jurisdiction even though the open burning is conducted in compliance with this Section:

(a) Fires purposely set for the instruction and training of public and industrial fire-fighting personnel.

(b) Fires purposely set to agricultural lands for disease and pest control and other accepted agricultural or wildlife management practices acceptable to the Board of Water and Air Resources.

(c) Fires purposely set to forest lands for forest management practices acceptable to the Division of Forestry and the Board of Water and Air Resources.

(d) Fires purposely set in rural areas for rights-of-way maintenance only in instances where there are no other practicable or feasible methods of disposal and under conditions acceptable to the Board of Water and Air Resources.

(e) Camp fires and fires used solely for outdoor cooking and other recreational purposes or for ceremonial occasions or for human warmth and comfort.

(f) Open burning of leaves, tree branches, or yard trimmings originating on the premises of private residences and burned on those premises in areas where no public pickup facilities are available and such is done between 8:00 A.M. and 6:00 P.M., and does not create a nuisance.

(g) Open burning in other than predominantly residential areas for the purpose of land clearing or right-of-way maintenance. This will be exempt only if the following conditions are met:

1/ Prevailing winds at the time of burning must be away from any city or town or built-up area, the ambient air of which may be significantly affected by smoke, fly-ash, or other air contaminants from the burning;

2/ The location of the burning must be at least 1,000 feet from any dwelling located in a predominantly residential area other than a dwelling or structure located on the property on which the burning is conducted;

3/ The amount of dirt or the material being burned must be minimized;

4/ Heavy oils, asphaltic materials, items containing natural or synthetic rubber or any materials other than plant growth may not be burned;

5/ Initial burning may generally be commenced only between hours of 9:00 A.M. and 3:00 P.M., and no combustible material may be added to the fire between 3:00 P.M. of one day and 9:00 A.M. of the following day, except that under favorable meteorological conditions deviations from the above stated hours of burning may be granted by the air pollution control agency having jurisdiction. It shall be the responsibility of the owner or operator of the open burning operation to obtain written approval for burning during periods other than those specified above.

(h) Fires for the disposal of dangerous materials where there is no alternative method of disposal and burning is conducted in accordance with procedures acceptable to the Board of Water and Air Resources.

(i) Permission granted by the Board under this section shall be subject to continued review and may be withdrawn at any time.

1.4 The effective date of the regulation shall be from and after _____ _____.

AMBIENT AIR QUALITY STANDARDS

1.0 purpose

It is the purpose of the following ambient air quality standards to establish certain maximum limits on parameters of air quality considered desirable for the preservation and enhancement of the quality of the State's air resources. Furthermore, it shall be the objective of the Board, consistent with the State Air Pollution Control Law, to prevent significant deterioration in ambient air quality in any substantial portion of the state where existing air quality is better than the standards. An atmosphere in which these standards are not exceeded should provide for the protection of the public health, plant and animal life, and property.

Ground level concentration of pollutants will be determined by sampling at fixed locations in areas beyond the premises on which a source is located. The standards are applicable at each such sampling location in the state.

1.10 sulfur dioxide

The ambient air quality standards for sulfur oxides measured as sulfur dioxide are:

(a) 60 micrograms per cubic meter annual arithmetic mean.

(b) 260 micrograms per cubic meter maximum 24-hour concentration not to be exceeded more than once per year.

(c) 1,300 micrograms per cubic meter maximum 3-hour concentration not to be exceeded more than once per year.

1.11   sampling and analysis

Sampling and analysis shall be in accordance with procedures published on April 30, 1971, in the Federal Register, Vol. 36, No. 84.

[Similar paragraphs for other pollutants—e.g., particulates, carbon monoxide, photochemical oxidants, hydrocarbons, and nitrogen dioxide would include state standards at least as stringent as those set by the national standards (Table 13–2).]

Maximum permissible emission standards   This is a class of regulation that prohibits quantities of contaminants from a specific source in excess of standards specified. It is, therefore, based on analysis of a source to determine if an allowable rate of discharge is being exceeded.

The application of the Ringelmann standards for control and prohibition of visible emissions is one example. Another example is control and prohibition of particulate emissions from a hot mix asphalt plant to ensure that specific aggregate process rates will not produce particulates in excess of maximum allowable emissions in a given period of time.

Examples of state air pollution regulations based on maximum permissible emission standards are

REGULATION NO. 2—CONTROL AND PROHIBITION OF VISIBLE EMISSIONS

2.0   purpose

The intent of this regulation is to promulgate rules pertaining to the prevention, abatement, and control of emissions generated as a result of fuel burning operations and other industrial processes where an emission can be reasonably expected to occur.

2.1   scope

This regulation shall apply to all fuel burning installations and such other processes as may cause a visible emission incident to the conduct of their operations.

2.2   restrictions applicable to existing installations

No person shall cause, suffer, allow, or permit emissions from any installation which are:

(1) Of a shade or density darker than that designated as no. 2 on the Ringelmann Chart for an aggregate of more than 5 minutes in any one hour or more than 20 minutes in any 24-hour period or

(2) Of such opacity as to obscure an observer's view to a degree greater than does smoke described in paragraph 2.2, subparagraph 1.

(3) All existing sources shall be in compliance with the provisions of paragraph 2.3 within 5 years.

2.3   restrictions applicable to new installations

No person shall cause, suffer, allow, or permit emissions from any installation which are:

(1) Of a shade or density darker than that designated as no. 1 on the Ringelmann Chart for an aggregate of more than 5 minutes in any one hour or more than 20 minutes in any 24-hour period, or

(2) Of such opacity as to obscure an observer's view to a degree greater than does smoke described in paragraph 2.3, subparagraph 1.

2.4

Where the presence of uncombined water is the only reason for failure of an emission to meet the limitations of paragraphs 2.2 and 2.3 those requirements shall not apply.

2.5

The effective date of this regulation shall be from and after _____.

REGULATION NO. 6—COMPLIANCE WITH EMISSION CONTROL STANDARDS

6.0   purpose

The purpose of this regulation is to assure orderly compliance with emission control standards.

6.1   scope

This regulation shall apply to all air contaminant sources both combustion and noncombustion.

6.2

After the effective date of any emission control standard, all sources of air contamination shall register with the Department of Water and Air Resources in accordance with the provisions of Regulation No. 4.

6.3

In determining compliance with emission control standards, means shall be provided by the owner to allow periodic sampling and measuring of emission rates, including necessary ports, scaffolding and power to operate sampling equipment; and upon the request of the Department of Water and Air Resources, data on rates of emissions shall be supplied by the owner.

6.4

Testing to determine compliance shall be in accordance with methods approved by the Board of Water and Air Resources.

6.5

All existing sources of emissions shall comply with applicable regulations and standards at the earliest possible date with all sources being in compliance within three (3) years from the approval of the State's Implementation Plan by the Federal Government. All new sources shall be in compliance prior to commencing operations.

6.6

This regulation shall be effective from and after its adoption.

EMISSION CONTROL STANDARDS

1.00   purpose

It is the purpose of the following emission control standards to establish maximum limits on the rate of emission of air contaminants into the atmosphere. All sources shall be provided with the maximum feasible control.

1.10   control and prohibition of particulate matter emissions from fuel-burning sources

No person shall cause, suffer, allow, or permit particulate matter caused by the combustion of a fuel to be discharged from any stack or chimney into the atmosphere in excess of the hourly rate set forth in the following table:

Heat Input 1,000,000 Btu/h	Maximum Allowable Emission of Particulate Matter (lb/h/1,000,000 Btu)
Up to and including 10	0.60
100	0.33
1,000	0.18
10,000 and greater	0.10

For heat input between any two consecutive heat inputs stated in the preceding table, maximum allowable emissions of particulate matter may be graphed proportionally.

1. For the purpose hereof, this standard applies to installations in which fuel is burned for the purpose of producing heat or power by indirect heat transfer. Fuels include those such as coal, coke, lignite, and fuel oil but do not include wood or refuse. When any products or by-products of a manufacturing process are burned, the same maximum emissions limitations shall apply.

2. For purpose of this standard, the heat input shall be the aggregate heat content of all fuel whose products of combustion pass through a stack or stacks. The total heat input of all fuel-burning units on a plant or premises shall be used for determining the maximum allowable amount of particulate matter which may be emitted.

Regulation of use or design of equipment   This type of rule regulates the equipment or process constituting the source of pollution to accomplish the desired reductions in the emission of air contaminants. This may be accomplished by banning the use of certain types of equipment, by establishing design standards for usable equipment, by establishing operational standards, or by requiring use of specific types of control equipment or control techniques. Compliance may be obtained by a requirement that all sources register and furnish sufficient information upon which to determine if the source should be allowed to continue operations. A license to operate may then be granted.

An example of an air pollution regulation based on use or design of equipment is

Use of Hand-Fired Equipment Prohibited

a. general

1/ This Section shall apply to fuel-burning equipment including, but not limited to, furnaces, heating and cooking stoves and hot water furnaces and heaters, in which fuel is manually introduced directly into the combustion chamber. It shall not apply to wood-burning stoves in dwellings nor to fires used for recreational purposes nor to fires used solely for the preparation of food by barbecuing.

b. prohibition

1/ After three years from the effective date of this ordinance no person shall operate or cause to be operated any hand-fired fuel burning equipment in the City of _____.

2/ The Commissioner may order that any hand-fired fuel-burning equipment not be used at any time earlier than set forth in this section whenever such equipment has been found to be in violation of the restriction of visible air contaminants contained in Section _____ on three or more occasions in any six-month period.

Regulation of fuels or fuel consumption   This type of rule might require use of smokeless grades of fuel or desulfurized fuel or prohibit use of specific types of fuel during certain seasons of the year.

An example of an air pollution regulation based on fuels or fuel consumption is

RULE 62—SULFUR CONTENT OF FUELS

This rule bans the use of high-sulfur fuel oils seven months each calendar year. The rule bans gaseous fuels containing sulfur compounds in excess of 50 grains per 100 cubic feet of gaseous fuel (calculated as hydrogen sulfide at standard conditions) or any liquid fuel or solid fuel having a sulfur content in excess of 0.5 percent by weight. This rule is in effect beginning with April 15 and ending with November 15, a period in which natural gas is in supply.

ORGANIZATION OF THE AIR POLLUTION CONTROL AGENCY

Most states have found it necessary to enact special acts to control air pollution and to provide a special agency to enforce these acts. In some cases, the antipollution agency handles both water and air resource problems. Some states encourage local cities, counties, or county groups to form air pollution control agencies and delegate authority to these local agencies for enforcement of state air pollution or local air pollution ordinances. Certain specific powers are usually retained by the states (e.g., jurisdiction over public utility power plants).

These local air pollution control agencies are responsible for keeping the public informed about air pollution programs and providing information to planning boards for consideration in zoning to ensure that air use is coupled with land use. These organizations should be properly funded and furnished qualified personnel to operate the continuous-monitoring programs, perform the source–emission inventory, and perform field control operations.

To achieve the objectives of the air pollution control program, it becomes necessary to reach *all* sources of air pollution in the field in order to effect their control. This can be accomplished by a field inspection and enforcement program. Field control operations are intended to determine air pollution potentials and solve air pollution problems. Since some of these problems may have no existing legal solutions, they may require special analytical testing and research in order to prove a violation.

field control operations

In practice, the field control operations break down into two distinct types of activities. The *first phase* is the patrol and inspection operation, which is concerned directly with contacting the sources of air pollution. Control operations may be predominantly concerned with preventive control through persuasion by means of imparting information, appealing to civic pride, and otherwise motivating voluntary compliance. The *second phase* constitutes the investigation and prosecution of violators. This phase is extremely important because the success or failure of all enforcement action depends on the accuracy and completeness of the technical data reported.

Patrol and inspection personnel are trained to detect visible violations from fixed or moving sources and serve notice of violations when they are observed, to maintain surveillance of all sources of air pollution, and to answer all specific source complaints made by citizens. After all the sources are properly registered, a continual source inspection is made of these plants to determine potential pollution and compliance or noncompliance with air pollution regulations.

Like other law enforcement officers, field patrols may operate by sectors in large cities. Within a sector, source inspections may be made on a scheduled basis or may be unscheduled. Central radio communications may be used to promptly check citizen reports of complaints. Aerial inspections provide means of detecting open burning and evaluating plumes from smokestacks.

Investigation and prosecution personnel gather evidence and witnesses for court cases and hearings, prepare specific charges, file complaints, and assist in the prosecution of violators. (A film, "Role of the Witness," listed at the end of this chapter, shows an example of some problems in this area.)

Because the control program requires registration of sources of air pollution, source testing, and evaluation of all technical source data, *engineering personnel* may be required, especially in the community that has a large industrial complex.

In a comprehensive agency, a *research unit* consisting of personnel from various scientific disciplines may be required to gather data, take samples, forecast pollution conditions, do laboratory analysis, and reduce and evaluate data.

Many variables involving the level of government and the complexity of the community make it impossible to suggest the organization of an air pollution agency in more specific terms than those given. Each agency at the state level or lower must formulate an organization in terms of specific needs and budgetary allocations.

SUMMATION

In this textbook the reader is confronted with this challenge— "Shall we surrender to our surroundings or shall we make our peace with nature and begin to make reparations for the damage we have done to our air, to our land, and to our water?"

There is a tremendous pride in being first or best in whatever you choose as a profession. We as a nation have been rightfully proud of the advances in technology that have brought bounty to our land and wealth beyond the wildest dreams of many other nations of the world. With technological achievement we have also brought destruction to our nation—yes, to our planet.

It is the hope of the author that readers of this book will not wring their hands in despair or join the carping crowds who seek to stop all future progress because of the danger of pollution. Rather it is hoped that readers will strive to attain more knowledge about pollution problems and how to use technological skills to control these problems in such a way that future progress can still be attained without the danger of ultimate destruction of the environment.

Air quality criteria, air quality standards, emission standards, and air pollution control regulations are the stepping stones to eventually repairing the damage to our environment. However, the same pioneer spirit that led to the conquest of this land must now be as arduously applied to the technological conquest of the pollution that threatens to destroy us.

REFERENCES

Law and Contemporary Problems, Vol. 33, No. 2. Durham, N.C.: School of Law, Duke University, Spring 1968.

State of North Carolina, *Rules and Regulations Covering the Control of Air Pollution*. Raleigh, N.C.: Board of Water and Air Resources, Jan. 21, 1972.

U.S. Department of Health, Education, and Welfare, *Air Pollution Control Field Operations Manual*, PHS, Pub. No. 937. Washington, D.C.: U.S. Government Printing Office, 1962.

U.S. Department of Health, Education, and Welfare, *Air Quality Criteria Pamphlet*, PHS (NAPCA), Pub. Nos.
   AP-49, Particulates, 1969.
   AP-50, Sulfur Oxides, 1969.
   AP-62, Carbon Monoxide, 1970.
   AP-63, Photochemical Oxidants, 1970.
   AP-64, Hydrocarbons, 1970.
Washington, D.C.: U.S. Government Printing Office.

U.S. Department of Health, Education, and Welfare, *An Air Resources Management Plan for the Nashville Metropolitan Area*, PHS, Pub. No. 999-AP-18. Washington, D.C.: U.S. Government Printing Office, 1965.

U.S Department of Health, Education, and Welfare, *A Compilation of Selected Air Pollution Emission Control Regulations and Ordinances*, PHS, Pub. No. 999-AP-43. Washington, D.C.: U.S. Government Printing Office, 1968.

U.S. Department of Health, Education, and Welfare, *Guidelines for the Development of Air Quality Standards and Implementation Plans*, PHS (NAPCA). Washington, D.C.: U.S. Government Printing Office, May 1969.

U.S. Environmental Protection Agency, *Air Quality Criteria Pamphlet* (Nitrogen Oxides), Air Pollution Control Office, Pub. No. AP-84. Washington, D.C.: U.S. Government Printing Office, 1971.

RECOMMENDED FILMS

TF-101   Role of the Witness (90 min)
           AQ Improvement Program (15 min)
           Available: Distribution Branch
                          National Audio-Visual Center (GSA)
                          Washington, D.C. 20409

QUESTIONS

1/ What is the purpose of the national air-sampling network?

2/ What is the basis upon which air quality control regions are designated?

3/ What federal agency has primary responsibility for air pollution in the nation?

4/ What is meant by air quality criteria? By whom are they established? What must be considered in developing air quality criteria?

5/ What is meant by an air quality standard? What is the purpose of air quality standards? What is the difference between air quality standards and emission control standards?

6/ For what six air pollutants has the federal government prescribed national air quality standards? When were they established?

7/ What unit of measure is now used to express air quality standards?

8/ What is the difference between annual arithmetic mean and annual geometric mean?

9/ Why is CO measured in mg/m^3 rather than $\mu g/m^3$? What is the air quality standard for CO?

10/ Why are all measurements corrected to a standard temperature and pressure? What is that standard?

11/ On what date were states required to submit air pollution plans?

12/ What are some of the things a state air pollution plan should include?

13/ What is the purpose of an air pollution control regulation?

14/ What should an air pollution control regulation encompass?

15/ What is the basis of an air pollution control regulation?

16/ What are the objectives of an air resources management program?

17/ Since the EPA has established air quality criteria and several air quality standards and emission standards, why does a local agency need an air quality monitoring program?

18/ What are three methods that may be used to register air pollution sources?

19/ What are the two types of law upon which legal aspects of air pollution regulations may be based? Explain the difference.

20/ What are four categories of regulations that may be established under statute law? Give an example of each.

21/ What may the methods of penalty be under criminal code?

22/ What may the methods of penalty be under civil injunction?

23/ What are some duties of patrol and inspection personnel?

24/ What are some duties of investigation and prosecution personnel?

PROBLEMS

1/ Have students do original research on the following subjects:

a/ Find out the organization of local or state air pollution agencies.

b/ Determine what system of air pollution control is authorized in the local community.

c/ Determine what air pollution laws are in effect to regulate mobile sources and how they are being enforced.

d/ Attend an actual courtroom trial pertaining to an air pollution case.

e/ Spend a day with a field patrol and write a report on the day's actions.

f/ Interview lawyers, judges, and men on the street to determine how many people are familiar with local air pollution regulations in the community.

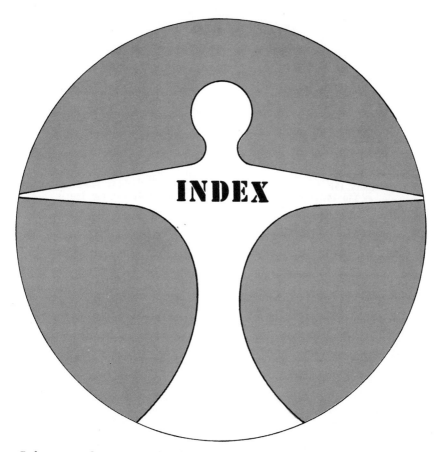

INDEX

References to figures are in **boldface**; references to tables are in *italic*.